Technologies of Refuge and Displacement

Technologies of Refuge and Displacement

Rethinking Digital Divides

Linda Leung

LEXINGTON BOOKS
Lanham • Boulder • New York • London

Published by Lexington Books
An imprint of The Rowman and Littlefield Publishing Group, Inc.
4501 Forbes Boulevard, Suite 200, Lanham, Maryland 20706
www.rowman.com

Unit A, Whitacre Mews, 26-34 Stannary Street, London SE11 4AB

British Library Cataloguing in Publication Information Available

Library of Congress Cataloging-in-Publication Data

Library of Congress Control Number: 2018936924

ISBN: 978-1-4985-0002-9 (cloth: alk. paper)
ISBN: 978-1-4985-0003-6 (electronic)

™ The paper used in this publication meets the minimum requirements of American National Standard for Information Sciences—Permanence of Paper for Printed Library Materials, ANSI/NISO Z39.48-1992

Printed in the United States of America

Contents

PART I: INTRODUCTION **1**

1. What's Technology Got to Do with Refugees? 3

2. Background and Methodology 11

3. Digital Divides: A Review of Literature 17

PART II: DIGITAL DICHOTOMIES **23**

4. Netizens and Asylum Seekers as Cultural Citizens 25

5. Technological and Social Determinism 35

PART III: ALTERNATIVE MODELS **47**

6. The Strength of Weak Ties 49

7. Actor Network Theory 59

8. Hierarchies of Technology Literacy 65

PART IV: PRACTICES AND PRINCIPLES **91**

9. Accessibility: Moving Beyond the Disability Paradigm 93

10. User-Centered Design 107

11. UCD Principles in Practice 121

Conclusion 125

Bibliography 127

Index 139

About the Author 141

List of Tables

Table 8.1. *Mind the Gap* research study respondents: key demographic information 66

Table 8.2. Comparison of Australian refugee experience with various technologies: before and after arrival (Leung, Finney Lamb, and Emrys 2009) 67

Table 9.1. Factors in minority group experiences with technology 96

Figures

Figure 8.1. Hierarchy of refugee telecommunications literacies (Leung 2011) 67

Figure 8.2. Base level telecommunications literacies with which respondents left their country of origin (Leung 2011) 68

Figure 8.3. Additional literacy of independent communication technology use acquired by some respondents while in their intermediate country (Leung 2011) 73

Figure 10.1. Ethra 109

Figure 10.2. Fahima 110

Figure 10.3. Aye 112

Figure 10.4. Prototype resource kits being tested (Leung 2011) 118

Figure 10.5. Digital literacies are constituted by a range of literacies. 120

Preface

The notion of "refugee" is complex because of its specificity within the remit of the United Nation's (UN) Refugee Agency, as well as its generality in also referring to asylum seekers, detainees, and those who have resettled as humanitarian migrants and become permanent residents or citizens of a new country. It is in this general context that the book uses the term "refugee," such that it could be a person who is newly displaced, a person living in a refugee camp, a person waiting in an immigration detention center, or a person living and working in an intermediate country or a country of resettlement. Considering this breadth and diversity, there is the common factor in the "refugee" experience of forced displacement and migration, whether this is past or current. The empirical data used in the book was collected from "refugees" in this broad sense of the term.

In technical terms, however, those who have been recognized by the United Nations High Commissioner for Refugees (UNHCR) as "refugees" are different to asylum seekers, people who have not yet had their claim to be a refugee processed. This distinction is critical to how they are treated by governments, and is acknowledged in the sections of the book that discuss immigration detention environments.

Part I

INTRODUCTION

Chapter One

What's Technology Got to Do with Refugees?

With some important exceptions to be discussed, an examination of the literature across a number of disciplines highlights the comparatively little attention that refugees have received as and in relation to other technology users, despite the critical importance of technology in the lives of refugees. There is growing consideration of the role of technology in, for example, sustaining connections between displaced family members where contact is tenuous and at risk of being lost. Such studies are critical in building a body of work that raises awareness of technology as a fundamental human right and therefore a basic necessity to which refugees should have access. In following this line of argument, other questions arise: which technologies are the most appropriate for refugees in camps and displacement settings? What are the practicalities of deploying such technologies on the ground? What inequalities emerge surrounding access to these technological tools and services? What are the repercussions of having limited or no access to such technologies for refugees who are displaced and those who have resettled?

This chapter begins, firstly, by attempting to understand the salient aspects of refugee experiences, accounting for their breadth and diversity.

That refugees are not represented or considered as users of technology is telling in the ways that technology is researched. Firstly, studies of technology largely focus on the new and on early adoption. Therefore, those who are not early adopters of new technology are overlooked and not regarded as technology users. Subsequently, groups and communities (of which refugees are only one) who are deemed outside the frame of reference of technology users, are identified as "have-nots" or on the "wrong" side of the "digital divide."

This book seeks to deconstruct these taken-for-granted perspectives on both refugees and technology. Using empirical data from surveys and interviews with refugees, the book will show refugees to be technology users, although they may not conform to popular and pervasive notions of what it means to be on the "right" side of the "digital divide," that is, to own and use a computer with mouse and keyboard in order to access the Internet.

As refugees and other groups do not fit this mold, the book demonstrates the "digital divide" to be conceptually and theoretically inadequate for understanding how and why technologies are used or not used. Indeed, the "digital divide" is illustrative of the dichotomous models used in Technology Studies (or STS).

Binary approaches to thinking about technology and how people use technology are unpacked in part II of the book, where notions of digital citizenship are examined: while considered an informal and alternative type of citizenship, chapter 4 "Netizens and Asylum Seekers as Cultural Citizens" explores how it privileges certain groups while excluding others, namely asylum seekers.

In the subsequent chapter, other dichotomies are explored. In Socio-Technical Studies (STS), ideas and trends surrounding technology are said to be either technologically or socially determined. That is, either technology has an inherent power that compels humans to use it (with success or failure dependent on the capacity of the technology), or technology adoption is entirely at the mercy of human decision making leading some to fail while other technologies succeed. Related to social and technological determinism, another approach to interpreting technological philosophies and practices is through the lenses of utopianism or dystopianism. In other words, technologies are often discussed in terms of their positive or negative impact. Upbeat discourses center on life-changing technology solutions to a problem, while pessimistic outlooks point to bleak futures where humans are enslaved by technology.

It may be that these dualities are simply a reflection of the on/off nature of binary code. However, once you try and fit humans into these dichotomies, it becomes apparent that such models work better for some groups than others. Refugees' experiences of technology help to dispel these theoretical divides, and force a rethinking of these concepts. Therefore, each chapter draws upon survey and interview responses from refugees to interrogate and deconstruct the dichotomies discussed above. The process of collecting such a large body of empirical data about refugees' relationships to technology is detailed in the following chapter.

Part III of the book applies the data to alternative models of technology use and access, the data being used to both test and illustrate Granovetter's theory of weak ties (in chapter 6), and Actor Network Theory (in chapter 7). The former chapter attempts to apply Granovetter's theory, which was originally conceived as a theory of social relationships, to the connections between refugees and their loved ones that are made vulnerable through displacement. The latter chapter offers a framework for understanding the acute significance of technologies, both old and new, to refugees. We see Actor Network Theory in practice as refugees struggle to sustain contact with families and loved ones while displaced.

Analyzing refugee experiences of technology through the empirical data also leads us to understand that accessing and using technology requires more than just technical or digital literacy. Chapter 8 "Hierarchies of Technology Literacy" discusses the other various literacies that influence and impact the uptake of technologies.

Finally, once the inadequacy of dualistic STS theories has been addressed and some conceptual alternatives have been provided, part IV of the book offers some principles-based practices for designing technology access and services in ways that are inclusive. Those concerned with providing services to refugees need to draw from different disciplines in order to make interventions that improve access to technology for refugees. These disciplines might include design and technology, but can also include approaches that have been used by other groups. Chapter 9 "Moving Accessibility Beyond the Disability Paradigm" looks at the principles and practices of making technology more accessible. Although this has largely pertained to the disability sector, it has broader relevance to other groups such as refugees, as it is underpinned by an ethic of inclusion that directly contrasts with the exclusion inherent in "digital divides."

Using refugees as a case study, chapter 10 "User-Centered Design" examines how service providers can combat technological determinism by developing and tailoring a product or service to the particular needs of the people who will use it. This means understanding the targeted user group intimately and removing constraints that impede access and inhibit use by them. Once a product or service has been designed, it should be tested with the target group. The chapter details how this can be done, even in the field, in situations of displacement.

WHO DO WE MEAN BY "REFUGEES"?

According to the United Nations High Commissioner for Refugees (UNHCR 2016), there were an estimated sixty million people forcibly displaced in 2015 due to war, conflict, and persecution. Deemed as "persons of concern," this includes refugees, asylum seekers, internally displaced persons (IDPs), stateless persons, and others of concern to the UNHCR.

Countries who are signatories to the Refugee Convention have an international obligation to provide protection to those seeking asylum. The 1951 Refugee Convention defines "refugee" as a person who has:

> well founded fear of persecution for reasons of race, religion, nationality, membership of a particular group or political opinion, is outside the country of his nationality and is unable, or owing to such fear, is unwilling to avail himself of the protection of that country; or who, not having a nationality and being outside of the country of his former habitual residence, is unable, or owing to such fear, is unwilling to return to it. (UNHCR 2010: 14)

While the Convention's definition of a refugee is limited to the experience of persecution and its formal recognition in order to be identified as a "refugee," the UNHCR's remit extends beyond this to other "persons of concern." These other "persons of concern" include people subject to forced migration due to war or natural disaster, in addition to those who are seeking asylum but whose refugee status has not been established. In this book, the term "refugee" is used in this broader context of "people of concern" who have been involuntarily displaced within or from their country of origin.

Australia's refugee program is inclusive of those who have been subject to persecution in their home country, as well as those who are subject to substantial discrimination, particularly where there is a gross violation of human rights. The humanitarian migration program has been a feature of Australian immigration for many decades, and has been a pathway for people who have been determined to be "refugees" by the UNHCR to enter and settle in Australia. However, those who have been recognized by the UNHCR as "refugees" are treated differently than asylum seekers, people who have not yet had their claim to be a refugee processed. The latter are subject to indefinite mandatory detention until their "refugee" status is established.

The "refugees" whose experiences constitute the empirical data used in this book include those who were officially recognized and allowed to settle, as well as those who were asylum seekers at the time. They may have been, at some point, newly displaced, a person living in a refugee camp, a person waiting in an immigration detention center, or a person living and working in an intermediate country or a country of resettlement. Within this breadth and diversity, there is the common factor of forced displacement and migration whether this was in the recent past or some years prior.

Why has refugees' relationships to technology been largely ignored until recently?

Although there has been increasing examination of how refugees appropriate technology to maintain connections with their virtual communities, this growing body of work has focused on those whose official status as "refugee" has been determined and who are living in resettlement countries. There has been less attention given to the appropriation of technology in the contexts of displacement and forced migration by refugees themselves and not agencies, when "refugee" status is indeterminate and there are limitations to access. Part of this work also includes studies of how technology is being used to control and manage asylum seekers. It is to this latter under-studied area of displacement and immigration detention contexts—where refugee voices are rarely heard (Fiske 2016a: 3)—that the book seeks to make a contribution.

The diversity of "refugee" experiences of technology, which can span from countries of origin through to intermediate and resettlement countries, has been explored in seminal special issues of key academic journals. In the *Journal of Refugee Studies*' special issue on information and communication technologies (ICTs) and globalization, editors Wilding and Gifford (2013) bring together a series of works that demonstrate how technologies are being used in the healthcare of refugees (Phillips 2013), by Karen youth as part of their resettlement in Australia (Gifford and Wilding 2013) and by Bosnians who have resettled in Austria (Hailovich 2013).

Other studies that concentrate on the use of technology by refugees living in resettlement countries, include: Kabbar and Crump's (2006) examination of the adoption of ICTs by refugee immigrants in New Zealand; McIver Jr. and Prokosch's (2002) exploration of how various technologies are used for information-seeking by immigrants and refugees in the United States; De Leeuw and Rydin (2007), Robertson et al. (2016) and Lems et al.'s (2016) research on the ways refugee youth represent their cultural identities in the creation of their own media productions. Research that has focused on specific technologies includes: Howard and Owens's (2002) study of the Internet as a medium for communicating health information to refugee groups; Luster et al.'s (2009) analysis of the critical importance of the telephone in reconnecting Sudanese refugees in the United States with their lost families in Africa; Glazebrook's (2004) study of mobile phone use among Hazara refugees on Temporary Protection Visas in Australia; Riak's (2005) research on how kinship rights of Dinka refugees are enacted through the telephone; and Charmarkeh's (2013) examination of social media usage by Somali refugees in France. Such studies explore how technologies are used where access to and literacies in those technologies is assumed to be unproblematic and does not fundamentally affect communication practices.

By contrast, an issue of *Refuge* with a special focus on technology emphasizes deployment in displacement contexts. Anderson (2013) evaluates the UNHCR's Community Technology Access program in refugee camps, while Danielson (2013) assesses how technology is used to sustain contact between the UNHCR office in Cairo and asylum seekers who have fled to Egypt as an intermediate country. In addition, there is Harney's (2013) work on the mobile phone as a tool of safety for asylum seekers in Italy negotiating the precarity of their immigration status. A subset of the research looking at technology in contexts of displacement and unofficial "refugee" status is its use in the administration and control of asylum seekers. This has been studied much more scarcely, but includes Ajana's (2013, 2013a) work on biometric profiling of asylum seekers in the UK, and Briskman's (2013, 2008) account of technology and surveillance in Australia's immigration detention centers.

Whether exploring resettlement or displacement contexts, the literature on refugees and technology can be situated within a larger body of research on technology and migration. However, one of the key limitations of this body of work is that the availability, access, and affordability of ICTs is assumed, as it is user groups for whom these are not problematic that have been studied (see Mitra 1997; Gajjala 1999; Mallapragada 2000; Melkote and Liu 2000). The use of the Internet (Kadende-Kaiser 2000; Graham and Khosravi 2002; Karim 2003; Parham 2004; Bernal 2006), phone cards (Vertovec 2004; Wilding 2006) and mobile phones (Horst 2006) by transnational migrants has been investigated, but has been skewed towards early adopters and the socioeconomically advantaged for whom there are abundant choices in relation to available ICTs, and neither availability, access, nor affordability are impediments. Madianou and Miller (2012) describe this environment of numerous communication opportunities as one of "polymedia," in which the preconditions of availability, affordability, and literacy are already met. The type of migration examined in these studies is different to and arguably more advantaged than the forced migration of refugees. As Robertson et al (2016: 221) maintain, "the refugee family is thus a particular form of transnational family, and one that has thus received little attention" (Robertson et al. 2016: 221).

That refugees have been comparatively overlooked as technology users may also be a symptom of the kinds of technologies that tend to be studied. There has been extensive study of how technologies, particularly ICTs, are adopted and used. However, much attention has been given to the rise of computing and Internet usage, rather than with telecommunications or other types of technologies. Disciplines such as Media and Cultural Studies take a broad view, where technology is considered the tool by which marginalized communities negotiate their social, economic, and cultural conditions (see Halleck 1991; Hall 1998; Cunningham 2001). Examples include Paul Gilroy' s (1993) work on the black Atlantic, which notes that books and records have been vital in carrying oppositional ideologies and philosophies across the black diaspora. Likewise, black independent film is often regarded as appealing to and mobilizing a black diaspora through the rejection of commercial cinema, which does not serve black communities (Diawara 1993: 6; Reid 1993: 5). Urban black youth have also been studied extensively in terms of their appropriation of dance and music technologies to overcome their socioeconomic disadvantage through the transformation of objects of consumption (such as the turntable) into new modes of production (Baker Jr. 1991; Gilroy 1993; Williams 2001). Within Asian diasporas, the use of cable and satellite, the exchange of video letters and taped Bollywood movies have been interpreted as forms of localized challenges to the centralized power of the broadcast media industries (Gillespie 1995; Ang 1996). While such studies of technology consumption has concentrated on groups other than refugees, the approach of Media and Cultural Studies highlights the breadth of technologies beyond computers and the Internet, which can be considered ICTs.

The phone, whether mobile or landline, is the key technology for refugees. Yet given their status as mobile populations, refugees' uses of mobile phones has had hardly any attention in comparison to Western youth. Technology Studies, it would seem, is preoccupied with the new and affluent: much is known about how mobile smartphones are appropriated by young people as a lifestyle item (Nathan and Zeitzer 2013; Lenhart 2010; Walsh et al. 2008). Similarly, there has been much interest in the use of social media as part of networked youth culture (Loader et al. 2014; McLeod 2012; Graham 2014), but despite a few studies of its use in emergency situations (Morris et al. 2014; Alexander 2014) research on its use by refugees during displacement and humanitarian crises are few. Likewise, the reasons why older technologies such as landline phones continue to be used, and the communities who prefer them, has not been explored in any detail.

Overall, the study of communities and communication practices that surround particular technologies has concentrated on groups other than refugees. A review of literature across disciplines such as Refugee Studies, Technology Studies, Media and Cultural Studies, and Global Studies has shown the study of:

- technology use by refugees has had less investigation than other user groups;

- the familial and diasporic networks of transnational migrants has included refugees less than other migrant groups;
- communities and communication practices that surround particular technologies has concentrated on groups other than refugees.

The next chapter details how the large and longitudinal body of empirical data used in this book regarding refugees' technology use was collected over a number of projects and research contexts.

Chapter Two

Background and Methodology

It is important to remember that while refugee communities and contexts differ, it is also necessary to look comparatively and at the "big picture" of technology use, access and provision in order to develop appropriate standards and policies that ensure at least a minimum level of availability and service. Currently, refugee experiences of technology are not only diverse and disparate, but also largely interpreted through service providers. That is, there is a dearth of data that is primarily sourced from refugees themselves about their technology use. As a way of addressing this, I am making anonymized data that I have collected from over one hundred surveys and interviews with refugees about their experiences of technology, publicly available for re/interpretation and analysis at http://trr.digimatter.com.

This online database is possibly the most comprehensive collection of primary data on refugees' technology use across various contexts of displacement, detention, and settlement. In sharing this data, I am encouraging interdisciplinary collaboration between students, scholars, and the fields of Refugee Studies and Technology Studies.

The data was collected over a number of years and projects in Australia, and so reflects the refugee groups and communities who were seeking asylum in and migrating to Australia on humanitarian visas over this period of time. Some of these groups are also typically represented in other Australian-based research studies of refugees, such as ethnic Hazara from Afghanistan (Glazebrook 2004), ethnic Karen from Burma (Robertson et al. 2016; Lems et al. 2016) and Sudanese refugees (O'Mara and Harris 2014).

In 2004, I became involved in refugee advocacy groups who were coordinating visitor programs to immigration detention centers. This inspired a pilot study of the impact of Australia's official policy of mandatory detention on how refugees maintained links to their diasporas. As a sociologist of technology, my interest was in how differences in technology-mediated communication occurred in the restrictive environment of immigration detention, compared with other contexts of forced migration, flight and displacement. Therefore, in addition to those who had experienced immigration detention, the research was expanded to investigate technology use by refugees from countries of origin, through flight and displacement, to countries of settlement.

The research questions for the pilot study, which was later published as *Technology's Refuge*, included: How are communication technologies used in countries of origin, during forced migration and in the settlement process? How are their benefits and limitations perceived? How are relationships of power surrounding these technologies negotiated? What, if any, virtual communities surround these technologies? How does technology assist refugees in sustaining connections with their virtual communities?

The pilot study analyzed a total of seventy-three interviews and surveys with refugees collected in 2007 and 2008 about their use of communication technologies across different contexts of displacement, detention, and refugee camps. Participants included:

- male and female refugees or asylum seekers;
- participants from different regions of the world, including Africa, the Balkans, Asia, and the Middle East;
- refugees resettled in the Australian community;
- former asylum seekers who had been detained within immigration detention centers;
- adults as well as those who arrived as child refugees.

Participants were selected to illustrate the broad range of refugee experiences and perspectives of communication technologies, before, during, and after displacement from their home country. Participants were recruited from refugee support networks and communities in Sydney, using a snowballing sampling strategy.

Surveys were distributed to participants who initially did not elect to do an interview. However, of the forty-three survey respondents, nineteen went on to give more in-depth insights into their technology use during displacement by granting an interview. The country from which the most survey respondents originated (fifteen in total) was Afghanistan, although fifteen other countries of origin were represented.

Twenty-seven interviews were conducted, of which two were in mixed group settings with men and women. In total, fifteen females and fifteen males were interviewed. Interviewees originated from the Middle East (13), Asia (10), Africa (6) and the Balkans (1). All interviews were either extensively noted, or recorded and transcribed. Most interviews were conducted face-to-face; in some instances they took place over the telephone.

Transcripts contained a mixture of stories about the use of communication technologies and participants' perspectives on their use. Reflective field notes were added to the data to aid interpretation. The analysis was conducted in two stages. Initially, I summarized each interview in terms of significant events, experiences, and stories before proceeding to coding and analysis. A coding framework for emergent themes was developed, before final analysis and writing up of results.

In November 2009, following the publication of the findings of the pilot study (Leung, Finney Lamb, and Emrys 2009), I conducted a community-based workshop on refugees and communication technology in Sydney. The aim of the workshop was to test and verify the findings from the aforementioned *Technology's Refuge* pilot study, and identify strategies for supporting refugees' use of communication technologies in displacement settings and during resettlement in Australia.

The workshop provided an opportunity for refugee communities, advocacy groups, international non-government organizations (NGOs), resettlement services and researchers to come together to discuss potential solutions for refugees. It focused on the experience of refugees from eastern and western Africa.

There were three central questions:

1. How can we help refugees communicate during war, in flight, and in refugee camps?
2. How can we help recently arrived refugees in Australia use new communication technology?
3. How can we help refugees in Australia communicate with family overseas?

To present these questions to the participants, three "voice maps," visual illustrations, and discussion points on A2-sized posters, were created prior to the workshops. The voice maps listed communication challenges relevant to each of the three central questions that had been identified from the *Technology's Refuge* pilot study. Quotes illustrating the challenges were also included. Down the left side were a series of questions or "talk-process" steps to further guide the discussions. The "talk process" was informed by principles documented in World Café conversation guidelines (World Café Community Foundation 2015).

An account of the workshop discussion and outcomes, based on comprehensive notes taken during the small group and plenary sessions as well as a list of recommendations and thoughts for future action was published as *Refugees and Communication Technology* (Leung and Finney Lamb 2010).

Both the pilot study and workshop highlighted refugees' greater familiarity with phone and telecommunications technologies over computers. This resulted in a one-year research project and report, *Mind the Gap*, funded by the Australian Communications Consumer Action Network, which focused particularly on refugees' telecommunications consumption in Australia. The project sought to:

- examine refugees' knowledge of telecommunications products and services when newly arrived in Australia;
- investigate telecommunications literacy in refugee settlement service provision; and
- develop a telecommunications consumer education program tailored to recent arrivals from refugee backgrounds.

Sixty-eight respondents were interviewed, with thirty interviews being analysed for the report (Leung 2011). Respondents had arrived in Australia as recently as 2009, and as early as 1995. While the Internet was not a ubiquitous technology for earlier arrivals, this was also the case for many who arrived later. Furthermore, the challenges of accessing the most basic communication technologies were common to both the earlier and more recently arrived. This research data was collected over 2010 and 2011 across four states of Australia.

Allowing for low levels of English and native language literacy, participants were surveyed and interviewed in person rather than in writing. In some cases, interviews were conducted in pairs to allow participants to translate questions and responses for one another.

There were three main groups of interviewees, grouped by regions of origin. These included respondents from:

- Iraq, Iran, and Afghanistan
- Sudan and other African countries
- Burma, Cambodia, and Thailand.

Respondents from Iraq, Iran, and Afghanistan included young people and older men over forty years of age. Corresponding with this diverse range of ages, the older men had been in Australia for over eleven years, while the young people had arrived within the last two years.

Respondents from the Sudan were almost all women in their twenties and thirties. Most had arrived in Australia from 2004.

Almost all respondents from Cambodia, Burma, and Thailand were the newest arrivals compared with the other groups, generally settling in

Australia from 2007. A broad range of users was also represented, from late teens through to respondents in their twenties, thirties and forties.

Furthermore, there was a miscellaneous group of older refugees and a settlement worker (a migrant himself) from outside of the three key regions.

There were also later interviews and surveys conducted that were not included in the *Mind the Gap* report, but will be referred to in this book. In total, the data collected spans fieldwork diaries dating back to 2004 through to surveys and interviews conducted up to 2011. Rather than capturing a snapshot of refugee experiences in a particular place and time, the book represents a longitudinal study of refugees' negotiation of changing technologies and landscapes during forced migration through to settlement.

Where possible, interviews were recorded, and if recorded, they were transcribed. As the transcripts show, interviews were generally conducted in English, which was perhaps more challenging for the transcribers than for the interviewees and interviewers. While transcribers had to contend with accents out of context, the interviewers and interviewees had the advantage of body language and non-verbal communication to interpret what was being asked or said by the other. Group interviews allowed participants to help each other negotiate meaning between researcher and interviewees. In the case of surveys, where participants may have had difficulty reading and writing in English, the researcher would ask the questions out loud and complete the survey on the participant's behalf.

Considering that all participants were surveyed or interviewed in Australia, there was a heavy reliance on the recall of memories and experiences from their countries of origin or intermediate countries, some of which were recent and others a number of years ago. Although the research is premised upon the authenticity of the participants' reminiscences (rather than observation or description of current scenarios), participants seemed to appreciate the opportunity to have a voice and tell their stories as this was not available to them when they were in refugee camps or immigration detention.

Chapter Three

Digital Divides: A Review of Literature

Having unpacked the notion of "refugee" and briefly discussed the burgeoning interest in their relationships with technology, it is now apt to begin to understand one of the most pervasive concepts in relation to technology access and adoption: the digital divide. In reviewing the literature in this area, the chapter examines how digital divides have been depicted in largely simplistic terms such as "haves" and "have nots." More often, there is reference to only one, as in the "digital divide," constituted by a right side and wrong side. Refugees are not the only group deemed as being on the "wrong" side. Other groups and communities who are regarded as excluded are:

- New migrants and refugees
- People from non-English speaking backgrounds
- Older people
- People of low socioeconomic status
- People in rural and remote areas
- Indigenous communities
- People with disabilities.

Altogether, the commonalities between these minority groups in relation to technology access and use can no longer be considered "minority" issues as they affect a significant proportion of the global population. In attempting to synthesize different bodies of literature relating to technology and inequality, the chapter finds that the use of technology by minority groups has been studied, but in isolation. That is, exclusion from technology use has been investigated within particular marginalized communities rather than across them. Thus, a more holistic approach to understanding differences in technology use and access may positively influence the design of services for refugees.

The chapter begins by interrogating claims made in relation to digital divides, particularly statistics that reinforce notions that digital divides consist of a large majority that is technologically competent and literate, while a tiny majority are laggards. On closer inspection, it becomes clear that this view is skewed toward the First World. Rather, refugees are part of a global majority for whom technology access and use is neither assumed or commonplace.

INTERROGATING THE STATISTICS

It is estimated by Barnard et al. (2013: 1715) that nearly 70 percent of the world's population are "digitally excluded" in the sense that they do not engage with digital products or services, or access the Internet. This is consistent with the 2015 ICT Development Index (ITU 2015) showing that two thirds of the developing world is still without Internet access. All African countries were found to have remained global ICT "followers" in that they were unable to "catch up" to developed countries to become ICT "leaders" over the five years from 2002 to 2007 (Ayanso et al. 2014). Furthermore, only 40 percent of the current world population has an Internet connection today (Internet Live Stats 2016). Underpinning much of the literature on the digitally excluded is a sense of deficit, that without Internet access, the group in question is deprived of a resource that others have:

> to lag in the use of technology is to remain behind a veil of limited knowledge and opportunities. (Selwyn 2004: 370)

Conversely, the International Telecommunication Union (2015) reports that 81 percent of populations in developed countries have Internet access. A recent report by the Australian Government (2015) claimed that 92 percent of Australians use the Internet. This suggests only a small minority are not using the Internet, but access once in the last six months qualifies as online participation regardless of frequency of use, level of digital literacy, or level of activity. With this in mind, it is likely that the number is much lower. But nevertheless, also underlying this study is an assumption that to be part of the 8 percent (probably more) who can't or don't use the Internet, or haven't done so in the six-month duration of the study, is to be socially excluded or disadvantaged.

It is clear that research into Internet access is uneven. On the one hand, the numbers would indicate that refugees are part of a majority of the global population that cannot or do not access the Internet and that this is symptomatic of the countries or regions from which they originate and to which they have been displaced. On the other hand, much of the research attention has concentrated on those who are technologically advantaged (especially young people, see Coombes 2009), with an accompanying narrative that particular demographic minorities are laggards. For example, Kluzer and Rissola (2009: 67) estimated that approximately 16 percent of the population of the European Union to be social excluded due to income poverty, low socioeconomic status and lack of employment: this population was also eight times more likely to be digitally disengaged and have lower digital literacies. Older people tended to be prominent in this statistic (Martinez-Pecino and Lera 2012: 876).

Refugees are represented in all these demographic groups: older refugees, those who do not have a stable income due to forced migration, those who have been educationally disadvantaged as a result of displacement, and those who come from developing countries where Internet access is a privilege. In this sense, they can be considered part of the global majority that cannot or do not participate online. But in studies of Internet use that have concentrated on the technologically well-connected (Buckingham 2007: 51), they are represented as a minority of laggards. Where this minority status intersects with others, issues of access become more pronounced: that is, online participation becomes particularly affected for those who have limited English language ability, limited technical literacy, and limited affordability to access ICTs (Vinson and Rawsthorne 2015; Migliorino 2011; Jung et al. 2010).

When considered collectively, the intersection of these minority identities cannot be regarded as marginal. Therefore, it is inappropriate to frame the "digital divide" debate in terms of a deficit hypothesis that posits that these groups and communities, and especially refugees, are in danger of missing out on the benefits of technological advances because they do not have the necessary literacies and finances to stay abreast of them. Such arguments are premised on a First World perspective that normalizes online participation, and disregards the multidimensional factors operating to exclude Internet access for the global majority.

SO MANY ON THE "WRONG" SIDE

Looking at online access statistically provides a clear picture of a majority/minority divide between those for whom Internet access is a privilege, and those for whom it is a given. This delineation between majority and minority can also be articulated respectively as a difference between developing and developed countries, whereby Internet availability and use is less commonplace in the former than in the latter. In this sense, there is a geographic definition of the digital divide.

This is often extended to differences in Internet access between urban centers compared with rural and remote areas. This literature argues that remote communities face greater challenges accessing online technologies due to factors such as a lack of infrastructure, high costs, and low community/government support (Townsend et al. 2013; Hale et al. 2010; Curtin, 2001). Such studies tend to focus on particular nations and their development of infrastructure to ensure availability of ICTs (Sein and Furuholt 2012; Suwamaru and Anderson 2012), but span the developed and developing world.

Issues of remoteness and technology access in countries of displacement were also borne out in the interview data. Refugees spoke about living in camps that were far away from towns and cities, and having to travel to urban or regional hubs to make phone calls:

> So I stay in Ghana in refugee camp—in very deep bush . . . You have to go and find water from far away. (A11)

> We are living in the region camp and this is very far from the capital city. (A25)

Previously, the digital divide has been simplistically defined in terms of those who do and those who do not adopt technology (Mitzner et al. 2010), or otherwise "haves" and "have nots" (Dewan and Riggins 2005). However, such definitions individualize much of the larger issues at work, which are essentially national and global (Bruno et al. 2011).

The tendency to frame digital divides as problems of the individual can be sourced to research that correlates online participation with levels of education and digital literacy. As Epstein et al. (2011) contend, debates about the skills needed for technology use move beyond the preoccupation with infrastructure for which governments are perceived as responsible, but then places the onus on individuals to become technologically literate. One stream of research on digital literacy takes the form of analyzing educational attainment and Internet use (van Deursen and van Dijk 2014), while the other evaluates programs designed to boost digital literacy (Huggins and Izushi 2002).

Studies of digital literacy have largely concentrated on inequalities in the developed world, examining factors such as home computer access and use (Vigdor et al. 2014). These factors are then associated with other socioeconomic and demographic variables, showing educational disparities between rich and poor directly impact levels of digital literacy. Those with low levels of education and income are the least likely to participate online (ibid.; Australian Government 2015). There are also age, gender, and ethnic dimensions to digital literacy with research studies finding that older people have less familiarity with, access, and confidence to use a range of technologies (Jung et al 2010), particularly older females (del Prete et al. 2011). Black and ethnic minority groups were more likely to access computers outside the home, adversely impacting their digital literacy (Ennis et al. 2012). Having English as a second language, or not having basic English skills was also considered an impediment to digital literacy (Eastin et al. 2015).

The policy response to these identified groups has been to develop digital literacy training targeted at particular communities (Broadbent and Papadopoulos 2013). For example, Huggin and Izushi (2002) assessed successful methods of stimulating ICT culture and skills development among those who live in rural areas. Jung et al. (2010) consider computer and Internet training programs for seniors. Del Prete et al. (2011) evaluate the effectiveness of a project in the European Union aimed at digital inclusion for older females. Kluzer and Rissola (2009) examine a range of e-inclusion initiatives for migrants and ethnic minorities across Europe. Macdonald and Clayton (2013) explore ways in which digital literacy can be enhanced among people with disabilities in England, while Naidoo and Raju (2012) look at the impact of digital literacy training on students in South Africa.

It is telling that almost all of the digital literacy training programs being studied have taken place in developed countries. In one sense, this reinforces the geographical dimensions of the digital divide mentioned earlier. In another sense, it suggests a degree of futility in delivering such training in the developing world, and particularly in contexts of refugee displacement and forced migration. This is because no amount of training can overcome the complex array of infrastructural, financial, educational, and linguistic requirements for online participation that are additional to the basic technical literacies needed. As Epstein et al. (2011) contend, solutions for addressing the digital divide in the First World oscillate simplistically between the provision of infrastructure (which is seen as the responsibility of governments) and the provision of training (which is seen as an individual responsibility to upskill) for the digital economy.

However, others acknowledge that it must go beyond making technology available through infrastructure and the provision of training. Sparks (2013) suggests a third area that must be addressed, the social and cultural factors that affect technology adoption and use, and lead to the gender, ethnicity, and age biases in online participation discussed above. Further again, others propose that there are a series of divides (Barzilai-Nahon 2006), while others prefer viewing it as a continuum (Warschauer 2002, 2003) or spectrum (Livingstone and Helsper 2007; Lenhart and Horrigan 2003).

In relation to refugees, it is clear that simple dichotomies are not appropriate. The issue of availability and infrastructure is a perennial challenge in countries of origin plagued by war and political instability, as well as in intermediate countries when residing in isolated refugee camps. Therefore, the provision of digital literacy training does not present a suitable fix, especially when refugee contexts and communities are also socioeconomically, educationally, and linguistically diverse. Because refugees do not easily fit models of digital divides that developed countries have invented for the First World, they are not considered as technology users or in the design of technology products and services. Refugees embody the exception to the pervasive and prosaic dichotomies of digital divides: Us and Them, inside and outside, safety and uncertainty, transience and permanence, citizen and noncitizen (Diken 2004).

The following chapters illustrate the binary oppositions that have been invoked when representing particular users of technology and technology itself. Users are either Netizens, or otherwise (like refugees) marginalized from technology. Technology is conceptualized as utopian or dystopian, and its design as either technologically determined or socially determined. Refugees represent a disruption to these models, which must then be transcended when thinking about how to best provide appropriate and tailored technology solutions and experiences during displacement and settlement.

Part II

DIGITAL DICHOTOMIES

Part II explores some key binary concepts applied to technology and its users, as they pertain to refugees. Following on from part I's overview of digital divides literature, chapter 4 explores how technology use is idealized and personified in the Netizen, while the asylum seeker represents its antithesis. A similar dichotomy exists in terms of how the effects of technology on society are perceived and interpreted. This duality is discussed in chapter 5, "Technological and Social Determinism": one position asserts that technology has major impacts on how we live that may have utopian or dystopian outcomes but which are ultimately unable to be controlled; the other position is that technology does not have a life of its own without being socially shaped and influenced. These particular dimensions of "digital divides" are examined in relation to refugee experiences, demonstrating that essentialist categories of inclusion and exclusion do not work in the interests of the marginalized and vulnerable, especially those who have been subject to forced migration and displacement.

Chapter Four

Netizens and Asylum Seekers as Cultural Citizens

In this chapter, specific reference is made to asylum seekers because it is precisely the lack of official status as a "refugee" that underpins their treatment by the State with suspicion. Nowhere is this more pronounced than in Australia, the only country that has a policy of mandatory and indefinite detention for all unlawful noncitizens (The Asylum Seeker and Refugee Law Project 2013; Fiske 2016: 19). Despite international criticism from the UN (Human Rights Law Centre 2015), Amnesty International (2016), Global Detention Project (2008), and Human Rights Watch (2016) for its practice of mandatory detention and offshore processing, Australia is influencing other countries' approaches to asylum seekers as we see the increasing use of immigration detention globally (Jakubowicz 2016; Harney 2013; Diken 2004: 94). As I write this, there are asylum seeker children detained on Nauru who are spending their fifth Christmas in immigration detention (Children Out of Immigration Detention 2017).

There are several commonalities between asylum seekers and how technology use and users are conceptualized. Interestingly, technology use is often framed as a kind of citizenship, with users considered as participating in a wider community. With this participation comes certain rights and responsibilities, but essentially it is about access and belonging. For example, the Netizen is a citizen of the online world through participation in Internet culture and communities. Conversely, those who do not use the Internet or have access to technologies that enable them to participate in online culture are deemed to be outliers who are on the "wrong" side of the "digital divide."

Similarly, questions of citizenship are also at the core of the plight of asylum seekers, where citizenship of their country of origin may be insecure, citizenship in the countries to which they have been displaced is out of the question, and the wait to be accepted as a potential citizen of a new country is fraught and uncertain.

This chapter attempts to juxtapose the mobile citizenships of both the Netizen and the asylum seeker, highlighting the former to be privileged in its freedom to information and its capacity to traverse national borders in order to access such information. The second is fraught with danger and precarity, and yet is deemed a basic human right. The chapter will also seek to further interrogate the similarities and differences between these citizen personas through the lens of cultural citizenship.

Notions of cultural citizenship have been recently debated as a possible alternative to formalized State–sanctioned national identity, and as a response to globalized and transnational identities. It is characterized by "the flexible nature of citizenship in terms of multiple loyalties that may transcend any particular state" (Delanty 2002: 64) and is embodied in different types of cultural citizens with diverse positionings. According to Delgado-Moreira (1997), the cultural citizen emerges from debates about whether citizenship is still meaningful in the era of post-national states, and how it might overcome the limitations of the State to embrace human rights in the context of globalization, transnational movements, and localism. Rosaldo (1994) argues that cultural citizenship offers subordinated communities a micropolitical platform for legitimizing demands in the struggle to enfranchise themselves. It is about claiming and expanding rights in the community through everyday conditions and participation in communities, neighborhoods, workplaces, churches, and activist groups. It is a bottom-up citizenship built by and for collectives.

The cultural citizen is one who is free to engage in what Appadurai notes as the "global production of locality" (1996: 188), to engage in a situated community, to bypass material notions of locality that are fixed to create ones that are relational, contextual, and fluid.

Both the Netizen and the asylum seeker enact cultural citizenship as described above, as they are characterized by their participation in cultures in the absence of State-endorsed citizenship. The concept of cultural citizenship attempts to address those traditional binary and exclusive definitions of citizenship by understanding citizenship through acts of agency and participation rather than as an official and privileged state of being that is given by governments. However, the limits of cultural citizenship become evident when applied to those who are stateless and/or displaced. While one type of cultural citizen, the Netizen, is celebrated as the quintessential cultural citizen, the asylum seeker finds their cultural rights and practices in relation to information and communication technologies (ICTs) curbed.

THE NETIZEN

How do we understand concepts of cultural citizenship as embodied in the Netizen? The Netizen is a citizen of technoculture, Net culture, and cultures of technology. The utopian moment of the Internet highlighted the democratic potential of online participation to engage a wider range of voices that may otherwise not be heard in the political process. It was argued that the Internet provided a flat and egalitarian platform for the Netizen to roam. The Netizen would be motivated to participate in their online community precisely because of its openness to a broad spectrum of representation. Thus optimistic discourses of freedom prevailed whereby the Internet liberated users from the constraints of geography.

However, in such idealistic discourses, issues of access were not considered problematic. All the State was required to do was make the necessary infrastructure available that would facilitate interconnected networks. In some cases, the State left this to the Market to make available. In self-selecting their participation online, the Netizen was also generally left by their respective States to regulate their own Internet activity.

As the poster child for cultural citizenship, the Netizen is deemed trustworthy enough not to require State intervention because she or he has volunteered their own resources to engage in digital citizenry. Thus, the Netizen is on the socioeconomically advantaged side of the "digital divide," one in which they are still the minority in many of their nation states. As at 2014, just over a quarter of the population in Africa and a third in Asia are Internet users, with the world average around 40 percent (International Telecommunications Union 2013, Miniwatts Marketing Group 2014). This means approximately 60 percent of the world's population is not using the Internet.

My argument here is that although the archetype of the Netizen is only representative of a minority, for the State it fulfils those expectations and parameters around cultural citizenship. As the darling of cultural citizenship, the Netizen poses minimal problems for the State if official citizenship is not open to question. That is, the Netizen is afforded his/her freedoms online only if their State-defined identity has been ratified.

The Netizen has become the model to which all other citizens should aspire because they invest their own resources to participate in their cultural citizenship. While the State or the Market made the Internet available to them, they pay service providers for their own access to the Internet, and they also purchase the technologies required for this access. In this sense, the model cultural citizen is not only socioeconomically and educationally advantaged, but is rewarded by the State for being so through the increasing transfer of government information and services online at the expense of other means of provision.

Nonetheless, the Netizen does have certain responsibilities in exchange for their online liberty. Those who do not hold official citizenship to the State are assumed to have questionable allegiances. That is, the freedoms afforded by cultural citizenship, the flexible and multiple points of reference, are considered dangerous in the hands of those who are stateless or fleeing their country of origin. With governments increasingly demanding allegiance to the State from its own citizens in exchange for free participation in cultural citizenship activities, noncitizens find their capacity to be good Netizens or cultural citizens under scrutiny because of their immigration status.

THE ASYLUM SEEKER

As another kind of cultural citizen, the asylum seeker shares many similarities with the Netizen. The asylum seeker is a global citizen who has taken the initiative of enacting their rights under international law to seek protection from war, famine, political instability, persecution, and natural disaster in their home country.

However, from the Australian State's point of view, it is this act of self-preservation by the asylum seeker that makes them undesirable cultural citizens. Australia's Migration Act 1958, section 189, states that any unlawful noncitizen, including any unauthorized arrivals who do not have a valid visa, must be detained until that person either obtains a visa or leaves Australia. In Australia, official and widespread misperception of asylum seekers as "queue jumpers" (MacCallum 2002) was instrumental in enabling the legislative changes requiring mandatory detention of persons arriving in Australia without a visa. Between 1992 and 1994, Australian law moved from permitting (but not enforcing) limited detention of asylum seekers, to a blanket policy of mandatory detention (HREOC 2004) that, at one point, had up to twelve thousand individuals in detention (Castan Centre for Human Rights Law 2003). Anyone who enters Australian territories claiming asylum is immediately and indefinitely detained either in an offshore or onshore immigration detention center (IDC) until their claims are verified.

The privilege of self-selection and regulation afforded by the Netizen is not accorded to the asylum seeker. While asylum seekers demonstrate initiative and resilience in leaving their country of origin, this is not celebrated by the State for a number of possible reasons:

1. Their State identity is contested, and with it their cultural citizenship rights.

2. They are not as resource-rich as Netizens: electronic communities typically involve the more educated and elite members of diasporic communities whereas the localities from which asylum seekers emerge, such as refugee camps, to quote Appadurai (1996: 193), "are the starkest examples of the conditions of uncertainty, poverty, displacement and despair under which locality can be produced. These are the extreme examples of neighborhoods that are context-produced."
3. Asylum seekers represent a different sort of diversity to Netizens, although both can be considered minority communities of sorts.

In this sense, asylum seekers articulate the struggles involved in producing locality, which according to Appadurai (1996: 189), include resisting the modern nation-state's attempts to define all neighborboods: asylum seekers impose themselves on these neighborhoods unexpectedly and without consulting the State first. Additionally, they represent the growing disjuncture between territory, subjectivity, and collective social movements: as they are stateless, or have abandoned their own State, their allegiances and affiliations are questioned.

The Australian nation state has sought to throttle any further attempts by asylum seekers to produce locality by creating their own space or to be part of any form of neighborhood, by limiting participation in virtual or online space. As Diken (2004: 84) asserts, the asylum seeker wants to "participate without identification" while the State wants them to "identify without participation," or in effect, to be the exception to the usual freedoms associated with cultural citizenship. This is manifest in varying degrees of restrictions, as at one stage, immigration detention centers were divided into closed and open detention sections. A report by the Australian Council of the Heads of Schools of Social Work (ACHSSW), noted that "many asylum seekers were held in separate "closed camps" and were not able to access telephones, newspapers, TV or mail for periods of up to 12 months" (ACHSSW 2006: 29). Thus, even participation in Andersonian notions of imagined communities (Anderson 2006) through media consumption was completely constrained. Items that were regarded as security risks by detention center management included crochet hooks, rollerblades, tape recorders, soccer boots, DVDs, wallets, nail polish, pencils, and pens (ACHSSW 2006: 33). Even use of the lowest of technologies (pencils and pens), and participation in those low-tech cultures (for example through letterwriting) were prohibited in closed detention.

In open detention, the technologies permitted have changed considerably over time and without any explicit policy regarding technology access. This can only be read as a form of regulation of cultural citizenship by the State, whereby participation in technocultures is highly controlled. During the course of the research, it was limited to low technologies such as pay phones, fax machines, photocopiers, television and video. Then mobile phones and some Internet access were allowed with restrictions. Regardless of the technologies that were permitted, the opportunities for the production of locality were severely restricted: just as detainees' physical movement was constrained, so was the potential to freely occupy or construct any kind of virtual locality. The rights of the asylum seeker as a cultural citizen were dealt with very differently to that of the Netizen, whose rights were a given, automatically bestowed with access to computers and the Internet.

Note that the technologies required to become a Netizen were not made available to asylum seekers in immigration detention until the beginning of 2007, and again with strict conditions. Detainees had a maximum of one hour per day, were not permitted to view blogs or discussion boards, nor "create" their own content (such as their own web pages). If detainees had their own computers, they could not pay for their own Internet access: modems were removed before the computer could be used. This provides an insight into the extent to which the State will discourage asylum seekers from engaging in any kind of cultural citizenship.

So then, how did asylum seekers practice their own forms of cultural citizenship under these conditions? Denied the rights of formal citizens while their claims for refugee status were decided by the State, asylum seekers demonstrated how informal, flexible modes of citizenship could be practiced while in immigration detention. However, this was contingent upon a number of factors that not only included the imposed institutional prohibitions, but also the cultural and financial capital of the asylum seekers themselves.

Firstly, the skills and literacies of the asylum seekers informed their capacity to participate. Whether they knew how to use a phone card, could use a pay phone, or read and understand English instructions affected the extent to which asylum seekers could engage in cultural citizenry. In other words, cultural capital was critical to the practice of cultural citizenship.

Secondly, irrespective of these cultural and technical literacies, access to technologies through which cultural citizenship could be enacted was severely restricted. In closed camps, participants spoke of not being allowed to read a newspaper.

> There was another guard in the detention centre and he was reading the Western Australian newspaper, and I questioned if he could lend me his newspaper to me when he finished it. He said, "No I can't give you my newspaper," so I said "Why?" And he said, "Well this is another rule here." I was quite sad. I think that I was visible in some way, that I feel sad and frustrated about his rejection. He came to me after about an hour and gave it to me and said in a quiet voice that you need to bring it back to me. And that was it. (Mr R)

In residential housing within the detention center compound, which supposedly carried with it more freedoms (your own TV, stereo, PlayStation, kitchen), there was also very little:

> No nothing. We don't have a gym, we don't have a library, we don't have an English classroom and we don't have a very big playground, like a basket ball or soccer, nothing, even the walking place, we don't have very big. Activities in here is nil. There's only one small one—ping pong ball. That's it. Five people, six people have been in there. How do you play? (Mr C)

Where technologies were available, participants suspected that access was sabotaged:

> I tell you, 200 detainees and all of them trying to make a phone call to their lawyer, family, friends, and there are four phones; two of them broken down. Most of the time, half of those phones are broken down. I was suspecting—although you could call me paranoid, but I was suspecting there is something very sinister going on because always two phones are broken down. I mean, this is very suspicious, that's very suspicious. I have no evidence that it's deliberate, but that looked very suspicious. (Mr U)

The interviewees did not comprehend the need for such severe restrictions, arguing that they wanted to participate in Australian culture, that they were curious to learn about the Australian way of life. Other asylum seekers insisted that such basic entitlements should not be taken away:

> cos it's a right to have—it's a right to community to be able, you know, to have access to the outside world. But before the introduction of the Internet and mobiles, we had very limited, no contact with Australian community . . . We should be given more access to normal life. More access to and links to the community. Because, look at it this way—we have hopes of living in the community. So to be able to integrate into society that you don't even know is very hard. So I think more links with the society sooner that would help in a way that if who go or leave would be able to adapt better. (Mr A)

Mr A argues that in the absence of official citizenship, cultural citizenship activities promote a sense of community, belonging, and social inclusion. As such, they are important and the State should not interfere with them. The restrictions were incomprehensible not only to the asylum seekers themselves, but to outsiders as well, with family members overseas unable to understand why making a simple phone call for detainees was problematic:

> I gave him a number but my mum didn't believe it. She said, why doesn't she ring me? What happened to her? She is in Australia—everybody has phones. (Ms Y)

It was also apparent that staff of the detention centers did not necessarily agree with the restrictions to asylum seekers' participation in cultural activities, with many aiding detainees in their cultural participation, if not actively ignoring its occurrence.

> Some officer they knows we had a mobile. They are very cooperate, they don't say anything . . . Most officer they ignore, they know, they say oh you have mobile. (Mr C)

In many cases, asylum seekers claimed their cultural citizenship rights in covert and haphazard ways, by working around the restrictions to access. In closed camps, they would crawl under the fence to make phone calls on the pay phone that was available in open detention. After three months in closed detention, Ms Y wanted to tell her mother that she was pregnant:

> We rang them and they didn't answer . . . I didn't know how to use this public phone because that's new for me . . . The phone was ringing, nobody was there. I was very angry. (Ms Y)

Asylum seekers would use the lowest of technologies available to them, so when there were long queues for the pay phones, and where they felt their conversations were being monitored or that they could not talk freely on the phone, they would write letters. These are examples of the insistent practice of cultural citizenship. Asylum seekers also demonstrated that where possible, any gains should be shared with others. As one of the few detainees who worked in the library, Mr C spoke about secretly making copies of the Migration Act available to others so that they could have some understanding of Australian immigration law when making their case for asylum:

> Before they don't allow us to read that book, they don't give us, but I work in the library. Somebody gave me then I pinch. Then I try make a copy then I give all the people, all the detainee. (Mr C)

Indeed, asylum seekers were often reliant on others both inside and outside detention for their access to cultural citizenship. Where websites were restricted inside the detention center, detainees would phone one of their contacts on the outside, ask them to print out the relevant pages and bring it with them the next time they visited. As recording devices (including any kind of camera, even those in mobile phones, as well as voice recorders) were not permitted inside detention, interviews with journalists and researchers like myself would be recorded over the phone.

But as with many other types of cultural activities, a third barrier to asylum seekers' participation was affordability, having the financial means to enact their cultural citizenship. Like Netizens, asylum seekers who had money had more access to technology than those who had none:

> I didn't have much money to buy a telephone card and you need to buy a telephone card—if you had money in your account, they could deduct from your account and then pay and you can buy a telephone card. I didn't have that as well. (Mr R)

When asylum seekers were in open detention and had access to pay phones but no money to buy a phone card, former detainees who had been released would buy a phone card and call their friends in detention to recite the access and PIN numbers to them. Likewise, before and after mobile phones were allowed in detention centers, former detainees and refugee advocates would pay for phone credit for those inside, ensuring that they had the means to phone or SMS family and friends on the outside:

> So what we did, we were lucky enough to smuggle some mobile phones inside detention . . . then I stopped working actually when I got a mobile phone and I got some visitors who were kind enough to pay for my prepaid credit sometimes . . . All I had to do, always hide my mobile, I keep it in vibrator mode . . . I think mobile phone is the heart of the communication for the detainees . . . Many times I call police, I had to call police, I had to deal with New South Wales Police because I call police for GSL officers or if there's a fight or if there's a criminal offence or something. (Mr U)

Asylum seekers who had experienced mandatory detention identified technologies as being key to their acculturation to life of Australia. Being denied access to such technologies while in detention made them more acutely aware of their importance once released:

> I didn't know how to use the Internet and I didn't know how to use a mobile . . . The first thing I bought [when I got out of detention in 2001] was a newspaper, because it was cheap and because I couldn't understand all the text. I had to know some of the meanings, so I needed to buy a dictionary, so I bought an Oxford Dictionary. It cost me a lot of money. I did buy a mobile phone a few days later . . . I bought a mobile, because a mobile was very important for me. It was good that I could keep in touch with my friends who were in Australia and through that, I could find a job, I had a number for people to call, so it was something that I needed. (Mr R)

The empirical evidence from asylum seekers demonstrates cultural citizenship to be highly dependent on networks. Networks enable asylum seekers to overcome obstacles imposed by the State in relation to participation in technology. Networks also allow asylum seekers to circumnavigate financial and literacy barriers to their cultural citizenship. Ultimately, asylum seekers' experiences detailed above depict the contrasting and contradictory responses by the State to two kinds of cultural citizen: the Netizen and the asylum seeker. We see that cultural citizenship is conditional upon socioeconomic status as the State endorses the privileged Netizen despite that this is a minority citizenry.

Appadurai (1996: 198) argues that there are three main factors that most directly affect the production of locality: the nation state, diasporic flows and electronically mediated communities. The Australian State, in particular, has demonstrated that it will intervene in the production of locality in cases where capital does not accompany those diasporic flows. Furthermore, participation in technocultures, in those electronically mediated communities, is restricted if State citizenship is open to question.

The Australian context of mandatory detention of asylum seekers illustrates how the State regulates citizenship officially through processes of immigration administration, but also culturally through restrictions on technology access and use. Unwritten policies allowing access to certain kinds of media while prohibiting others suggest that the liberty of being a Netizen (that is, a cultural citizen) is only available when one's citizenship is not under question by the State.

Although this chapter has been about how the asylum seeker negotiates State interventions in their cultural citizenship in the Australian context, it also raises questions around privilege and access to technology. In other displacement situations, such as refugee camps, how can cultural citizenship be encouraged and enabled? How can a sense of belonging and participation be fostered using appropriate technology platforms that do not entrench binary divisions between "haves" and "have-nots," or asylum seekers and Others.

Chapter Five

Technological and Social Determinism

In addition to the binary of technology user versus nonuser, there are two main perspectives used in the study of technology: one is that of *social determinism*, whereby technological innovation and change is regarded as socially, politically, culturally, and economically situated; the other is *technological determinism*, which considers technology to be the catalyst for social change.

Technological determinism is much criticized within socio-technical studies (STS) for its underlying assumptions. These include unfounded links between technology and "progress," or beliefs about technology having an inherent power to transform society for the better or worse (Henwood et al. 2000). At its most extreme, technological determinism can be seen in idealistic and utopian, as well as dark and dystopian, visions of technology futures. It offers simple yet magical answers to complex problems that often entail "cause and effect" equations: introduction of a technology = un/desirable social effect. For example, it can be seen in assertions that social media will revolutionize the experience of displacement by maintaining connections between family members during forced migration; or that biometric profiling of asylum seekers will aid in the governance of forced migrants (Ajana 2013). There are often unquestioned assumptions underpinning technologically deterministic ideas, one of the most insidious being that technology leads to socioeconomic progress, and so everyone must "keep up" with the technology revolution by having the latest device and the skills to use it or be "left behind." As Henwood et al. (ibid.) argue, the simplicity of technological determinism is the principal reason for its endurance, but it leaves little room for human agency.

Social determinism takes a constructivist approach, reversing the causality of technological determinism by positing that technology is socially shaped and therefore a dynamic site of political and cultural struggle between people, groups, classes and institutions. It interrogates the ways that ideas about technology make their way into policies such as ensuring every child has a laptop, or establishing high-speed broadband networks. Social determinist or constructivist views offer a critical counter to any tendency to "evangelize" new technologies, as well as a healthy skepticism of pessimistic generalizations about the impending doom a technology will wreak.

It is tempting to prefer one position over another, but as Ajana (2013: 582) argues, the aim should be to move beyond both technological and social determinism:

> In considering these technologies and processes, the aim is not to posit or adopt a somewhat technologically determinist approach to asylum policy and its critique, whereby the technological is seen as possessing the entire agency. Nor is the purpose to foreground a form of social determinism whereby technology is seen as being merely a tool responding to the needs emerging from policy debates and political strategies.

As the following examples demonstrate, it is difficult to take apart a technologically determined approach without a social determinist lens; and conversely, the latter is often not enough to undermine the power and pervasiveness of the former. Also, it is more than possible for neither to be human-centered, to be used in combination to act against humanity and in ways that perpetuate disadvantage.

A socially determined policy, such as Australia's system of mandatory detention of asylum seekers, is a case in point. Anyone who enters Australian territories purporting to be a refugee escaping from persecution, political instability, war, natural disaster, and famine in their home country is immediately and indefinitely detained in an immigration detention center (IDC) until their claims are verified. Technology is instrumental in the enforcement of such a policy, ranging from technologies of containment, surveillance, and identification used in immigration detention centers (IDCs); to the prohibited access to information and communication technologies (ICTs) to detained asylum seekers. Here we see social determinism work hand-in-hand with technological determinism, specifically to create a climate of fear among asylum seekers to discourage them from coming to Australia, as well as to generate fear of asylum seekers in the Australian public.

The technologically determined and deliberately dystopian environments of IDCs are illustrated by the strict regulation of detainees' use of technology, based on unwritten policies that allow access to certain kinds of media while prohibiting others. IDCs are designed to set up extreme dualisms between those who control policies and practices of technology use in immigration detention and those who are subject to or incarcerated by them. Technologies of surveillance generate fear among asylum seekers and those who dare to visit them. They determine and patrol the boundaries of containment, who is able to move between them, and act as signifiers of the consequences of unauthorized crossings. The policing of space through technology is regarded as vital in dealing with asylum seekers, the ultimate "space invaders." Strategic to this limiting of space is to make the IDC a "non-place," somewhere that is almost invisible in its remote location and untouchable as "contact with the outer world is physically minimized behind the fences, which yield no permission to touch the outer world, resulting in the complete isolation of the refugee from public life" (Diken 2004: 91). Inside IDCs, asylum seekers' claim for space (both national and physical) is restricted by limiting their access to technology. Such practices speak to the potential creation of space enabled by technology, a techno-utopian (but still technologically determined nonetheless) idea. The liberating possibilities of technology become a source of fear when available to those considered undeserving of such freedom. The nexus between asylum seekers as signifiers of space invasion, and technology as a signifier of space creation and delineation, demonstrate the binary divides that operate within technological determinism, between techno-dystopian perspectives (technology as policing mechanism) and utopian (technology as freedom from borders).

TECHNO-DYSTOPIA

What constitutes a technologically determined dystopia? How might it be defined and articulated? In one sense, it is the practical application of technology in ways that work against humanity (such as through control and surveillance). In another sense, there is also an affective function to the technology: to instill a sense of fear and apprehension in those who are subject to it. This can be described as its emotional design, which goes beyond utility. The emotional design of IDCs exhibit all the characteristics of negative affect such as anxiety, fear, and anger (Norman 2004: 29, 104). Such design strategies are also evident in other systems and technologies of immigration administration. Wilding and Gifford (2013: 500) contend that " . . . nation states are using an increasing array of ICTs in border management and control. ICTs have thus enhanced the ability of nation states to monitor, administer and control 'desirable' population movements such as those of tourists and skilled migrants, while at the same time extending the surveillance of the state to identify more easily, and deny entry to, unauthorized and so-called 'undesirable' migrants, including refugees, asylum seekers and others without documentation."

The Australian immigration system has been designed to make the seeking of asylum unattractive, risky, and opaque. It has been constructed as a technology of prevention and exclusion, as articulated through its policy and practice of mandatory and indefinite detention. Australia's response to asylum seekers arriving by boat is to intimidate them with military technology and force in a humanitarian crisis:

> Helicopters flew over every day and night frightening everyone. The SAS, all in black, did crazy manoeuvres in their rubber boats. (Captain of the Tampa cited in Briskman et al. 2008: 30)

> It was during the night. We didn't know which way the shooting was going but the shooting was too much . . . [I was] vomiting, very scary, very sick and my daughter too, my daughter very vomiting. (Briskman et al. 2008: 33)

In addition to systems of administering asylum seekers, there are technologies of representation which complement them, such as print, broadcast and online media. Indeed, media technologies have been instrumental in escalating moral panics surrounding the seeking of asylum (see Mares 2002; McMaster 2002; Kushner and Knox 1999) and in perpetuating the widespread misperception of refugees as "queue jumpers" (MacCallum 2002). In this sense, media technologies occupy as much a divisive role in terms of delineating between Us (Australians) and Them (asylum seekers and refugees), as the technologies of control and surveillance that are found in immigration detention centers.

Thus, the flipside of how these large systems of administration and representation are designed to engender fear and anxiety in the Australian context, is how they are actually experienced by asylum seekers, how they are negotiated at the interface, on the ground. There is clear evidence of the dehumanizing effects of mandatory detention from different stakeholders, from research studies other than my own (see Fiske 2016, 2016a), and at each stage of the asylum seeking process.

> As I was driving in the car on the way [to the IDC], I thought about all the technologies which keep me connected to others. I had my mobile phone by my side in case I had to contact someone, or if someone wanted to contact me. As long as I have this, I am part of a communication network. But I wouldn't be allowed to bring this in.
>
> I was glad to be going with [some more experienced visitors], as they offered to direct me there in a car convoy, and escort me through the procedures required to enter the world of immigration detention. They were my protective panopticon within a more oppressive one. They would also be my connection to the outside world once inside when my ties with other networks are relinquished.

The architectural manifestation of this severed connection is evident in the chain mesh fencing and razor wire, surrounding what would otherwise be a park with outdoor tables and seating—the physical limitation on free movement. The physical environment's influence on the psyche is clear when detainees do not come out to meet their visitors.

Their names are called out on a loud speaker, but they begin to limit their own contact with the outside world in response to the imposed restrictions.

I was stopped at the street entrance by the guard who spotted me carrying a mobile phone. I was instructed to leave it in the car. [My companion] was also told to leave her handbag in the car.

When we arrived at the shed, we attempted to complete our 3 forms with the 2 pens available, one of which didn't work. We proceeded through a guarded gate into the visitor's centre, each took a ticket and waited for our number to be called. We managed to coordinate who we would request to see, as detainees are only allowed into the visitor's courtyard if their name is called on the loud speaker.

I was "processed" first: I submitted my driver's licence, declared I had been before so they could retrieve my details from the database. They cross-check who I have requested to visit (in case they've been deported), and put an identifying wristband on me. The guard told us to collectively keep our stuff together in a locker, which had to be negotiated with a code that gave you your assigned locker, and a PIN of your choice to open the locker. Once this hurdle was overcome, we tackled the X-ray conveyor belt and the metal detector one by one, after which we were then stamped with invisible ink. The next stop was handing our forms onto another guard, who presses the button to release the door and yells the names of the visited detainees out over a loud speaker. The final stage is a guard who notes down the numbers on the wristbands before letting you into the compound.

The loudspeaker blares out their names, and it is only through the subsequent haphazard delays that one can only guess that they either haven't heard from the inner compound, or don't want to come out. Then begins the process of having someone go in to find and fetch them. (Extract from fieldwork diary, July 2005)

Other visitors verify the technologized and dystopian environment of IDCs:

We needed to provide at least one name of an inmate we wished to visit, and a maximum of four. Fabia wrote down four names on each of our admittance forms. We passed through security checks and metal detectors underneath barbed-wire fences and into a yard with benches and a visiting area. (Isaacs 2014: 3)

Lawyers have commented that:

> Conditions of detention are in many ways similar to prison conditions: detention centres are surrounded by impenetrable and closely guarded razor wire; detainees are under permanent supervision; if escorted outside the centre, they are, as a rule, handcuffed. (Burnside in Briskman et al. 2008: 15)

Visitors have also reported being denied entry to an IDC because of having the wrong shoes and other arbitrary but ever changing rules surrounding visits (Briskman et al. 2008: 273–274). With each visit came the discovery of a change, such as having to book a day in advance by either fax or post, or only being allowed to talk to the detainees you had booked to visit, all part of "a purposefully backward system, to make it difficult to see the men" (Isaacs 2014: 245–246).

While visitation procedures change constantly and vary across detention centers, technologies have been central to the ways in which asylum seekers have been administered. In the example above, technology fulfils a number of utilitarian purposes such as containment, surveillance, and identification. Perimeter technologies serve to isolate those who are detained from the outside world and vice versa. Security technologies determine who and what are allowed to pass through the boundaries delineating inside and outside. Verification technologies confirm those who are permitted to return to the outside world once they have ventured inside. These are all part of the operations of IDCs, which themselves are like "blackbox" technologies in that they are mysterious and "largely shrouded in official secrecy" (ACHSSW 2006: 6).

Such design strategies are intended for those who are incarcerated by these technologies and those who dare to disrupt the relationship between inside and outside by visiting detainees. There is a clear asymmetry between those who design, use and control the technology (immigration detention center staff) and those who are subject to their processes and effects (detainees and their visitors), inevitably leading to abuses of power. Alizadeh (in Bharat and Rundle 2012: 136) refers to the "arbitrary wickedness" of the guards and "callous sadistic bureaucrats." Humiliation and intimidation by guards in detention has been well documented, including surveillance of refugee advocates and any staff who were friendly with detainees (Briskman et al. 2008: 133, 271, 277, 282). This is also detailed in Isaacs' (2014: 78) account of offshore immigration detention in Nauru: "Over time, we began to realize that the Wilson guards monitored all people within the camp, not just the men. They monitored Salvos, listened to our conversations, recorded our interactions, and tried to catch us out."

Techno-dystopian commentators describe such a relationship as one of "master and slave," part of a deliberate campaign to maintain such a dichotomy and alienate those held captive within these technologized environments. This is recognized by the detainees themselves, one of whom said that there is "one Australia outside the razor wire, one inside" (Jaivin 2012: 202), pointing to the distinction between having agency over technology, and being subject to its control.

Access to particular technologies was not always completely deprived, but highly controlled. In the case of one IDC visited by the author, detainees had access to pay phones, faxes, photocopiers, television, video: technologies that may all be characterized as low-tech, analogue, "old media."

Visitors were allowed to give detainees phone cards so they could use the pay phones without charge or the need for change. Detainees could also "work" for their phone cards by accumulating points for the number of hours worked. As Isaacs (2014: 16) explains: "Phone cards can be purchased at the (Nauru) camp canteen run by Transfield. They use a points system. The system awards men a weekly quota of points that can be spent on anything within the canteen. Chocolate, MP3 players, phone cards and cigarettes. Supplies run out quickly and nothing ever seems to come in." Briskman (2013: 12) adds that "Although the obligation to work was discontinued, during a visit in 2012 to the newly opened Yongah Hill detention center in Northam, Western Australia, an advocate was told that . . . men could make calls out but no one could call in to them."

In addition to pay phones, detainees were provided with access to a fax and photocopier, which were generally used to liaise with and send relevant documentation to lawyers. Detainees distrusted using the fax machine at the detention center because it was in a management office area and they required permission to use it. It meant the guards could read the faxes that were sent, as well as those that were received before notifying the detainees that they had received one. Detainees also had television, videos, DVDs, and newspapers. There were computers available, and prior to Internet access, some detainees loaded computer games on them to play. Others had PlayStation in their rooms. This low-tech, resource-scarcity is verified in Isaacs' (2014: 21) later study of Nauru: "What few resources we had quickly deteriorated due to overuse. Cards went missing, board games were broken, volleyballs would go flat and DVDs would scratch. The frustrations of camp life would often boil over thanks to a missing card in a deck."

It was noteworthy that the only technology to which detainees had access and which facilitated real-time person-to-person interaction was the fixed line telephone. The phone offered the opportunity for direct contact without the visual and other sensory realities of detention (as photos and cameras have never been permitted). For informal interactions, reliance on the telephone was typical for most detainees who had families on the outside, whether local or abroad.

Because there was no explicit policy regarding access to technology while in immigration detention, there was inconsistency across detention centers in relation to what was permitted, and this also changed over time. As Briskman (2013: 10) notes, "There was a time when the ban on mobile phones was relaxed, only to be later reinstated." However, the technologies that were especially not allowed were those that were new, enabled mobility, and the creation of space. Just as detainees' physical movement was constrained, so was the potential to freely occupy or construct any kind of virtual environment. While use of fixed line telephones was permitted, mobile phones were not (and this is still the case according to recent reports from offshore detention centers). This points to what Harney (2013: 542) describes as the "space-adjusting" capacity of the mobile phone and its potential for "spatialized individuality" (ibid.: 544).

Computers, especially those which are networked, appeared to be a potent source of fear (Mulgan 1996: 1). Thus, computer rooms were established and administered by petty socially determined rules to deal with the overwhelming demand for too few computers (Isaacs 2014: 74):

> The computer room rostered the men's Internet usage. Six computers; thirty minutes at a time on a slow connection. The opportunism that ran rife in the camp was evident here. Take as much time as you could even if it wasn't your turn. If there was a free computer, jump on it. The camp mentality was survival of the fittest. (ibid.: 22)

Therefore, technology use was obstructed so that only low-tech passive consumption, rather than active experimentation, was allowed. Such technologies of virtual space creation were considered threatening in the hands of those who have physically made unauthorized border crossings. Those who cannot remain within their own perimeters were perceived as undeserving of such technology. Technology signifies a way of being (that is free, mobile, always accessible, and always able to access), but it also connotes an ideal type of user, one that is appropriate and worthy of such technology. It seems that asylum seekers are not entitled to space nor mobility and, therefore, do not have rights to media that is considered to facilitate these qualities, in spite of their detention.

What does this suggest about the perceived dangers of "new media" and the resonance of last century's techno-utopian discourses? Given that detainees were only given access to "old media," it seems that this tired but resolutely upbeat rhetoric about new technology that celebrates it as inherently liberating has actually inflected policies determining the kinds of technologies to which detainees have access. Fears that accompanied previous technologies—such as the personal stereo's ability to build an acoustic space from which users can retreat from everyday life (Cranny-Francis 2005: 71)—have been recycled and translated for newer technologies.

The possibility that large, reliable, secure technologies that police space can be undermined or contested by newer, mobile, adaptable technologies which create virtual spaces is, on the one hand, a fearful techno-dystopian prospect. Yet, on the other hand, the assumptions of freedom and democracy embedded in the latter are techno-utopian and implicit in policies that deny asylum seekers access to new technologies that allow them to move freely within virtual space to communicate with the world outside of the detention center. The "liberating" nature of such technology is regarded as unsafe in the hands of asylum seekers, whose freedom of movement is institutionally contained by the Australian government through mandatory detention. Thus, asylum seekers in immigration detention are subject to the conceptual legacies of techno-utopianism and offer a sobering reminder of their long-lasting and potentially detrimental effects.

Although techno-dystopian projects such as Australia's mandatory immigration detention system are explicitly inhumane, it is worth reiterating that they are fuelled by a fear of the possibilities of a techno-utopia: virtual environments that are free from borders and enable the creation of networks, spaces, communities and alliances. Furthermore, such notions of techno-utopia and dystopia can be socially constructed in ways that are anti-human. Socially determined approaches to technology do not necessarily work in the best interests of humanity.

TECHNO-UTOPIA

A direct contrast to the techno-dystopian environments of immigration detention can be seen in Kaspersen and Lindsey's (2016) vision of the role of technology in refugee futures:

> We may see the rise of online direct-giving platforms, where refugees can prove their identity and describe their current needs, and individual donors can electronically transfer money to them. We may see widespread use of drones to drop aid supplies. Such innovations could start to make the traditional methods of delivering humanitarian aid—large organizations with a sense of pride in always having workers on the ground, physically proximate to beneficiaries—look as anachronistic to donors as old-style taxi companies in the age of Uber.

Indeed, there are many examples of positive and well-meaning refugee technology projects that are premised upon a technological determinism that equates technology with improvement. Grounded in the notion of communication as a human right, as expressed in Article 19 of the Universal Declaration of Human Rights (United Nations 1948), which includes the "freedom to seek, receive and impart information and ideas through any media regardless of frontiers"; these initiatives interpret and enact the article in different ways, but are often concerned more with what the technology can do than with the specific needs of refugees.

Some projects read "freedom of information" as ensuring humanitarian aid agencies have the latest technologies with which to do their work. One example is Noula, an online platform for mapping locations where help was both needed and provided during the 2010 Haiti earthquake. The platform was developed so that disaster survivors could "crowdsource" assistance. That is, rather than waiting for aid, people could communicate what assistance was required while others could volunteer whatever services or help they could offer. However, rather than people using the platform directly, it was found that the local community wanted to call and talk to someone about what was needed. In other words, it was a system that had to be mediated because users were either not familiar or comfortable using it, or could not access it directly. It was found that low-tech systems such as local radio and SMS were actually more effective in communicating vital information, while Noula took months to gain traction: "The key lesson of Haiti for international responders is that for information and communication systems to deliver, they must engage local populations and their technical capacity as equal partners, and they must understand and connect with existing systems before developing new ones" (Wall 2011: 6).

Similar lessons were learned from an earlier system called Ushahidi, deployed in the Democratic Republic of Congo in 2008, which invited those living in conflict zones to report incidents of violence (Ruffer 2011: 25). In this case, people could submit a report directly to the system by logging onto the website or sending a text message. However, the conditions of displacement, people fleeing, no network availability, and fear of being identified through reporting, all worked against people willingly submitting information.

In other projects like Noula and Ushahidi, the particular needs of affected refugees are absent altogether. Rather, the focus is on the technology alone and its capacity to enable information to be safely managed for and between aid agencies to do their work. Examples include Villaveces's (2011: 8) Libya Crisis Map case study, Wolfinbarger and Wyndham's (2011: 20) study of geospatial technologies as visual evidence of displacement, and Hall's (2011: 10) discussion of the piloting of digital radio as a substitute for analogue radio in the Philippines. Currion (2011: 41) argues that the deployment of technology must shift to being more refugee-centered: "Most of our discussions still focus on how responding organizations can use technology more effectively, rather than how disaster-affected communities might use these same technologies."

Education initiatives in refugee camps are also sources of technological optimism that often bypass the needs of the students. The Jesuit Refugee Service rolled out the Borderless Higher Education for Refugees project, an online learning program with students in the Kakuma refugee camp in Kenya and Dzaleka refugee camp in Malawi. Each site had a computer lab with Internet access which had to be housed in a secure building to protect the equipment from dust and air conditioners to regulate the temperature. Even with the most willing students and instructors, the challenge was ensuring they had sufficient electricity to power the labs. Dankova and Giner (2011: 12) concluded that "the use of technology to bring tertiary education to refugee camps is not a solution to protracted refugee situations."

A similar conclusion was reached in an evaluation of the UNHCR's Community Technology Access (CTA) program, which sought to provide refugees with access to education through the provision of technology (Anderson 2013: 22). With the aim of inspiring independent learning and promoting self-reliance for refugees, the technology alone was not able to provide personalized skills assessments that were tailored to the needs of local labor markets, and so did not improve employment outcomes for refugees. There was a tendency to "dump hardware in schools and hope for magic to happen" (ibid.: 23).

The hope that a technology can solve complex problems surrounding access to information or simplify communication, can result in a "tragedy of failures and false expectations" (Danielson 2013: 34). Many proposed technology solutions for refugee situations are premised upon availability of and access to the Internet or mobile networks, such as systems that allow refugees to receive overseas remittances (Duale 2011; Omata 2011). However, Price and Richardson (2011: 15) contend that there are still many countries where Internet access is low (such as the 1 percent of the population of the Democratic Republic of Congo which has online access), and while mobile phone coverage is generally greater, in some countries like Burma / Myanmar, it is less than 1 percent of the population: "In many rural communities the preferred method of accessing information is through spoken communication—through public announcements, meetings, events, radio and TV broadcasts as well as face-to-face contact" (ibid.).

Techno-utopian projects only fulfil their ambition if availability, access, literacy (technical and language), financial means, and cultural norms are not, in any way, problematic. Indeed, the outcomes of many of the aforementioned initiatives have shown that older, low-tech, or more traditional media, lead to more effective outcomes (Danielson 2013: 35), because they take into account the human realities and social determinism of technology adoption and use.

Part III of the book examines alternative models for thinking about technology and its users that go beyond binary and dichotomized notions of "divides," that are more human-centered and better aligned to the needs and experiences of refugees.

Part III

ALTERNATIVE MODELS

The chapters in part III offer alternative ways of understanding refugees' relationships to technology that are not simplistically dichotomized, but rather flexible and adaptable to users in low-tech and resource-scarce environments. Both chapter 6 on Granovetter's thesis on the strength of weak ties, and chapter 7 on Actor Network Theory (ANT) emphasize the socially determined ways in which technologically mediated networks are used. However, ANT explicitly acknowledges the equal role that technology and humans plays in these networks, such that they are as much socially as technologically determined by human and nonhuman actors. ANT manages to tread this nonbinary fine line to perfectly articulate that it is neither one or the other, but rather a complex interdependency. These fluid, networked models are further extended in chapter 8 to construct technology literacies as part of a wide spectrum, rather than a "have" or "have not."

Chapter Six

The Strength of Weak Ties

Although originally conceived as a theory of social relationships, this chapter attempts to apply Granovetter's (1983) ideas to the connections between refugees and their loved ones, which are made vulnerable through displacement. Citing examples from refugees' experiences of displacement and immigration detention, the chapter argues that seeking asylum weakens strong ties with family members, such that refugees become almost totally reliant on weak ties for survival. Granovetter's theory of the strength of weak ties is used as a framework for understanding the ways in which technology is appropriated by refugees to maintain their tenuous and fragile relationships with the world outside refugee camps and detention centers.

Granovetter's findings span high and low density networks of people with strong and weak ties. His (ibid.: 201) hypothesis asserts that a network constituted by an individual and their acquaintances is one of low density and weak ties. Conversely, the same individual and their close family and friends comprise a densely knit network with strong ties. Much previous research in the areas of Family Studies (Madianou and Miller 2012), Migration and Diaspora Studies (Robertson et al. 2016), Global Studies and Inter-Cultural Studies (Baldassar et al. 2007) demonstrate technology to be a vital component in sustaining strong networks and ties between relatives. The examination of socio-technical networks of refugees in refugee camps and detention centers show that they are unique in that they rely almost exclusively on weak ties, as their strong ties have been eroded by displacement.

Granovetter's work has been concerned primarily with human (rather than technological) networks, and the ways in which weak ties and low-density networks facilitate the formation of strong ties and high-density networks. While this has been examined in relation to job-seeking strategies and how job hunters find information about prospective employment, it is also relevant to refugees whose high-density networks are compromised by displacement. This is further exacerbated in contexts of detention, creating conditions whereby asylum seekers are highly dependent on technology to strengthen their weak ties.

Using Granovetter's ideas in relation to refugees illustrates that their strong ties (to family, community, homeland) have been undermined if not severed altogether due to displacement and detention. Refugees' high-density networks are especially jeopardized in detention because of restrictions to their access to technology, establishing the conditions for weak ties to form: "extensive weak networks can remain viable only when close ties are prohibited" (Granovetter 1983: 222).

HOW STRONG TIES ARE UNDERMINED

In the case of immigration detention, as seen in the previous chapter, ties with anyone outside the detention center were discouraged by geographic isolation and the representation of asylum seekers as inhuman and dangerous (Diken 2004). In addition, for many detainees and those who had experienced long periods in refugee camps, close ties with family had been decimated by forced migration.

> But for me in our country we don't have water facility. So we draw water from the well, not from the town. So my mother sent me [to go with] my mother's friend. For I had to fetch water. So on that particular day we don't know that that day is going to be the day [that our country would be attacked]. So we are on that place at the water well. The men started shooting. So I don't know the whereabouts of my parents. Straight away that woman I follow. My mother's friend just run away with me. So we went to Guinea direct. So since then I never set eyes on my mother, brothers and sisters. (A18)

One of the interviewees (A11) recalls his last memory of Togo, his country of origin: he was shot in gunfire and fell unconscious before waking up in a hospital in Ghana.

Although he knew that some family members were killed, he did not know where those who survived had fled. For the next twenty years, he would live in a remote refugee camp in Ghana:

> So I stay in Ghana in refugee camp—in very deep bush. They really actually are a ghetto with the people still there, like a prison. It's really sad. (A11)

Another respondent (A10) talked about fleeing Liberia in 1996, and losing contact with his mother until 2009. It was only through networks of weak ties that he discovered the whereabouts of his mother:

> From another relative who had his uncle, his brother here (in the refugee camp). He come together with his brother and after he met me and asked his relative overseas whether he knew my mum's whereabouts and the person said yes. (A10)

He was reunited with his mother again over the phone. But this involved the distant relative having to tell the mother the number of someone in the refugee camp who had a mobile phone who was prepared to let the respondent receive the call.

With strong ties destroyed through displacement, "the life of the asylum seeker is marked by an extreme isolation; not only physically but also socioeconomically and culturally" (Diken 2004: 92).

HOW WEAK TIES DEVELOP

Given the tenuous state of their strong ties, refugees have to develop and rely upon their network of weak ties. However, according to Granovetter (1983: 209), weak ties are more difficult to form than strong ones. Many of the respondents originated from countries and communities where relationships were mainly constituted by strong ties, where families were in close proximity and there was little means to connect with others beyond the immediate vicinity. One of the interviewees (A27) claims that "90 per cent" of the population of his home country of Sudan was living without phones or mobiles. Others affirmed this by saying that even basic technology like electricity was unavailable in the areas they lived in their country of origin. When strong ties are decimated through forced migration, refugees have to learn whole new ways of relating to people who are not in their network of strong ties. Weak ties might include distant relatives or family members who no longer live nearby. Refugees living in camps waiting for resettlement were often reliant on those who had already been resettled in another country for support:

> Well, phone was really expensive. The only way you get a phone is they have someone overseas who send you money to get a phone. (A10)

> My family got one, yeah. My uncle, who was in Canada, and then he talked to mum and said that maybe you can buy phone so we can talk to each other. (A14)

> Especially Somali people, because some Somali people, their family are in America or maybe they can send the money and then they can buy and just use. (A33)

Refugees not only depended on familial weak ties to get access to a mobile phone, they would also then need the assistance from acquaintances to learn how to use it:

> Those are being using a phone before, they have to show you. The first thing you learn is how to okay the phone when a call come in, how to okay, and secondly how to dial the number on the phone. (A10)

In immigration detention centers, weak ties included visitors and the array of organizations (churches, refugee advocacy groups, law firms, health organizations) to which they belonged. The people and organisations that encompass this free-form socio-technical network are the synapses which mediate detainees' contact with the outside world. Although face-to-face interaction is a key part of the relationship between visitors and detainees, technology is also vital to the sustenance of these weak ties. Fiske (2016a: 41) asserts that detainees recognized the importance of weak ties and were strategic in formulating them: "Establishing direct lines of communication with members of the Australian community was a high priority for asylum seekers, both in terms of a strategy toward regaining rights and in terms of creating a political space in which their words and deeds were meaningful."

TECHNOLOGY'S ROLE IN WEAK TIES

It was also evident that technology was critical in sustaining and strengthening weak ties in refugee camps. An interviewee (A11) says that the mobile phone enabled him to find work in the refugee camp because he was easily contactable. It also enabled him to lobby embassies for resettlement. In other words, technology is key to allowing information to be transmitted within a network of weak ties. Despite that respondent A11 is unable to read or write, he was able to tap into information about potential employment by using technology to maintain communication with weak ties.

Weak ties are the means by which refugees receive information from within and beyond the confines of refugee camps and detention centers:

> individuals with few weak ties will be deprived of information from distant parts of the social system (ibid.: 202)

Such information is critical to refugees' understanding of the social, political, economic, and cultural context to which they have been displaced and how to negotiate its practices. While refugees are subject to systems that purposefully create alienation, Granovetter (1983: 203) argues that weak ties serve to counter this marginalization.

Given the geographic isolation of refugee camps and detention centers, the information disseminated through weak ties is crucial to community and network building.

> You cannot contact people in the camp because there's nobody in the camp. There's no telephone in the camp so you just—it's like, isolated, so there's nothing in the camp. (A19)

> Because I was staying in refugee camp, there's no public phone in refugee camp. (A33)

Without telephone services in refugee camps, access to technology becomes dependent on access to money or weak ties that can provide access to information. This can also be seen in the context of immigration detention, where systems are explicitly designed to weaken strong ties, and make communication through weak ties difficult. The strategy for doing this is by limiting access to technology as illustrated in the previous chapter.

In the Australian system of mandatory detention of asylum seekers, offshore processing of applications for asylum on Christmas Island, Manus Island, and Nauru are deliberate policy decisions intended to cut both strong and weak ties. The remoteness of the islands severely constrain physical travel to and from them for lawyers, media, and human rights organizations. The cost of travel to these islands from the Australian mainland is prohibitive. The lack of infrastructure on the islands means that the technology on which detainees depend to sustain connections with their strong and weak ties can be unreliable, and even if it does work, they must compete to use it assuming they are allowed to use it.

The introduction of and increase in offshore processing in 2001 and 2012 respectively (Isaacs 2014: xiii, 5, 184), in favor of detaining asylum seekers in immigration centers on the Australian mainland shows that restricting physical as well as technology access all but severs weak ties. In both cases of offshore processing and mainland detention centers, mobility is discouraged politically and physically through the restrictive environment in which detainees are imprisoned. Arguably, however, the mainland immigration detention centers, especially those in capital cities, have a greater capacity for weak ties to flourish simply because there is better access to detainees and a superior technology infrastructure to facilitate their interactions with the outside world. Weak ties become the tools by which refugees mobilize or practice a form of virtual mobility, and become key to their survival. Weak ties have a bridging function in terms of connecting differences and to social circles external to one's own.

In 2005, the Australian Migration Amendment (Detention Arrangements) Bill allowed detained families with children to live in community detention, that is, in residential accommodation outside of an immigration detention center. Families were informed in writing by the Department of Immigration that they were going to be released imminently into community detention although dates and times were not specified. As many were released before regular visitors to detention centers were able to see them again, the weak ties between the detainees and the visitors were temporarily severed, demonstrating the tenuous nature of such relationships. Nonetheless, these connections were reignited with phone calls from families to one of their visitors. The information was then passed by the visitor to other refugee advocates. This is an example of the haphazard but functional dynamic of a network of weak ties, and one that is far less likely to happen in offshore processing camps because of the lack of regular visitors.

Weak ties help to facilitate adaptation and change: for refugees, this relates to acculturating to the new environment to which they have been displaced, and navigating its particular practices, processes, and procedures. Granovetter (1983: 205) articulates this as "weak ties have a special role in a person's opportunity for mobility."

INFORMATION TRANSMISSION THROUGH WEAK TIES

In another example, a detained family had been notified in writing of their imminent deportation by the Department of Immigration. Some visitors to the detention center talked with the family about what could be done during the waiting process: one of them undertook to look up information online (when Internet access was not available in detention centers) about making a formal application under a section of the Migration Act in which the minister can exercise individual discretion on particular cases. The other visitor undertook to contact refugee advocacy groups for advice as to the legal options open to the family. Once that information was found, it was faxed to the family. Without these weak ties, networks become insulated and such information would not have been available to the family in question. Again, this probably could not have happened in an offshore processing center because the information systems are less robust and there are fewer weak ties available who can act as conduits for information.

Comparing the data between participants in immigration detention centers and those from refugee camps, there is evidence to suggest that there were greater opportunities within the latter for communities of weak ties to form. In part, this was due to the size and scale of some of the camps in which some of the participants had resided; and also the lack of punitive restrictions surrounding technology access. In other words, in refugee camps, there was freedom to be entrepreneurial and creative in the ways that weak ties could be deployed to open up the world to refugees. Those who could not read or write, or couldn't use a mobile phone, could resort to weak ties. If a letter needed to be sent, those living in refugee camps could ask a person traveling to a particular area to deliver it.

> When I want to write a letter I just call my friend who knew how to write, to write for me. (A11)

> Maybe if you write a letter, maybe someone might take it for you but not post. (A14)

> We send someone, like if I type a letter and someone I know knows where the person lives I will say oh can you take this letter to this person for me. You just send it by hand. Or in some days we never had stamps or just write that person's name on the address and just send it . . . that's the only way we can get to someone. That was the only way we could get to someone. If we don't do that we can't get to contact the person. (A29)

> If you have somebody you knew, then he can help you out. If you don't have somebody you know you may not have access to the radio. (A30)

> So I ask them. They sit with me like how you sit down. They say if you want call, go here. After calling, go this place, do this one, do this one. If you want make the contact, if you want to write the person and then the number, go here, go in. (A11)

However, this does not mean that weak ties rely only on goodwill. Weak ties could also involve business transactions.

> In Kenya I used to learn a computer. Yeah, I pay the money and I have a teacher. (A33)

Refugees in camps spoke of purchasing a mobile phone and getting taught how to use it from the business they bought it from:

> They used to sell mobile phones. So they have to teach you how to use it. That's how we learned this. (A30)

In other cases, weak ties were forged by working for an aid agency or the government, which meant access to technology:

> Unless the government, the [unclear] government, maybe you are one working there, so we call to the office telephone. But the normal people, they could not get it. You had to call the office. If there is no one working there, you cannot get them. I'm using telecommunication, telephone at the government office. You can go and pay the money, they can call the person if you know their phone number. (A31)

> I use [mobile phone in Kenyan refugee camp] every day because I was working with UNHCR. (A33)

This reliance on weak ties also suggests an imbalance of power, whereby refugees must depend on the kindness of others, or otherwise pay others for access to technology. Such asymmetric relationships made refugees vulnerable:

> It's only for the office. Also the government is controlled Muslim. He will not allow someone who is not Muslim sometimes to use it. Very big discrimination. If you are not Muslim you cannot get it. If you want to apply for that job, you cannot get it, you are not a Muslim. Very difficult. (A31)

> Because if you're in that office, you are post parcel, you can use. If you have friend there, maybe you can beg them because they're not allowed for refugee to use that kind of stuff. (A37)

> They send packages to family members left behind—not an easy task to accomplish. People living in camps have no addresses, so senders nest envelopes within envelopes to relay their packages through their transnational networks, making particular use of churches and their congregations. At each point of the package reading a church congregation, prayer group or church choir, an envelope is shed from the package revealing the next destination, until it eventually reaches its target recipient. (Robertson et al. 2016: 225)

Although strong ties are more readily available and the ones to which people immediately refer, weak ties are generally the only option for refugees as the conditions of displacement and immigration detention compromise their connections to family members.

In the absence of strong ties, weak ties are the only option for sustaining relationships with the world outside of refugee camps and detention centers. Technology plays a vital role in this process with refugees making use of every technology that is available to them to connect with people and the world beyond the boundaries of their displacement. It becomes clear through applying and extending Granovetter's hypothesis to refugees that weak ties are only as strong as the technologies that support them. This paves the way for thinking about technology as key components of refugee networks, and for moving toward a socio-technical understanding of relationships.

This is further discussed in the next chapter on Actor Network Theory.

Chapter Seven

Actor Network Theory

While Granovetter's theory of weak ties is about the strength of human-to-human connections, Actor Network Theory considers the relationship between humans and nonhumans. Instead of regarding this as a fixed binary, Actor Network Theory suggests that there is a fluid and symbiotic relationship between humans and nonhumans such that any action or agency is impossible without both (Law 1999: 3). Taking this as its departure point, Actor Network Theory is concerned with movement and mobility between actors (who may be human and nonhuman) and networks. Networks are constituted by more than the actors themselves, but by the circulation of and relations between actors (Underwood 2008: 1). Latour (1999: 223) suggests that "Actor Network Theory seeks to describe the actions of humans and non-humans symmetrically and as partners in interaction."

The scenarios discussed in the previous section demonstrate the critical importance of nonhuman elements in maintaining connections in refugee-visitor networks. Without letters, landlines, or mobile phones, relationships with both strong and weak ties with people could not have been sustained. Technology is an actor that can be described as a "portable extension of affective human relationships" (Ortico in Brady and Dyson 2014), but which is "part of a complex hybrid and heterogenous collective consisting of many actors" (Brady and Dyson 2014).

However, Actor Network Theory has also been criticized for elevating the value of the nonhuman. Suchman (ibid.) argues that nonhumans are inferior partners and that the scope of interaction between people and machines or objects is limited. Latour (1992) agrees that technologies both constrain and enhance human capabilities. This is apparent in cases where refugees have to "make do" with the technologies that are available and devise work-arounds for tasks that could be done much more quickly and effectively using other tools. For example, interviewees A48 and A50 note that as there were no phone lines to communicate with family members outside of their home town in Sudan, cassette tapes were used instead. For refugees who were unable to write letters and have them delivered to other towns, and who did not have access to phones, cassette tapes were used to record important information that needed to be communicated to family living far away. Where using a phone network could deliver that information far more quickly, refugees had to resort to more cumbersome means relying on more actors and slower networks.

Nevertheless, how might Actor Network Theory help refugees? For service providers working in displacement contexts such as refugee camps, Actor Network Theory is a potentially useful way of understanding how localized interactions and communication happen on the ground through various devices and processes, and how this is then translated to macro level forms. It looks at the way the "really big" (ibid.) relates to small sites and practices, and how networks are shaped and change shape accordingly. This can relate to how education, health, shelter, food, or water are provided to refugees; how service providers establish and maintain direct lines of communication with refugees; and how information is disseminated among refugees.

Actor Network Theory understands networks to be complex landscapes (Law 1999: 7). It invites us to examine how networks handle the activities that constitute it: how is agency facilitated or hindered? Who or what are the key actors bearing the stress and/or helping the network to function? Law (1999: 3) argues that the notion of the Actor Network is an intentional oxymoron as the relationship between agency and structure is interwoven and cannot be separated. They are two aspects of the same phenomenon, a hybrid of social and technical forces and forms (Bolter and Grusin 1999: 77).

The intrinsic link between agency and structure is clearly illustrated in the environments of immigration detention centers. In the Australian context, whether these centres are onshore or offshore, their structures use nonhuman actors to privilege one set of human actors over others. The human actors in the network are staff and detainees. However, the technologies used in the detention centers, for both communication and surveillance, enable agency for staff and stifle agency for detainees. For example, while staff have unlimited and free access to communication technologies, this is restricted for the detainees. Prior to 2007, detainees had access to pay phones inside the center but were not allowed mobile phones. One of the detainees said that he spoke to his wife and children on the center pay phone every few days. However, the call costs were expensive as his family only owned a mobile phone. For them to call him was also expensive and awkward, because they had to call the pay phone and if somebody answered, they had to then try and find him somewhere within the compound. This exemplifies the strategic deployment of a nonhuman actor (public pay phones that were unfamiliar to detainees and required financial resources to use them) by the Australian government to ensure that network agency is highly restricted.

Similarly inside detention centers up until 2007, computers were available to detainees but with no Internet access. Therefore, computers were deliberately removed of their network capabilities, while restricted access to networked technologies such as fax machines privileged access by staff over detainees.

Visitors to detention centers were instrumental in bringing "contraband" items such as phone cards, mobile phones, and documents (Briskman et al. 2008: 277), illustrating how actors can actively resist a closed network. Examples of how technology can also be used to work around networks that have been designed to limit agency include refugee advocacy groups using the media to mobilize supporters to visit detainees, and deploying email campaigns to raise funds to assist refugees (Briskman et al. 2008: 272). Arbitrary rules of the technologized detention network, such as not allowing cameras inside or within five hundred meters of an IDC (Fiske 2016: 73), were overcome through sophisticated means of smuggling. Newer, human-centered networks of resistance were established that took the form of strikes, sit-ins, rooftop protests, riots, complaints, legal actions, escapes, art work, theater (ibid. 64).

Interviewees demonstrated their network literacy by highlighting the differences in networks between their countries of origin, the countries to which they were displaced, and Australia as their country of settlement. A46 has no memory of using any technology in Sudan as a child, but in her intermediate country (Egypt), she recalls, "We had phone in our house. Home phone. My Dad had a mobile phone."

The home phone in Egypt would be used to call her friends, and the family would use it two or three times a week. Therefore, networks in her country of origin are portrayed with mostly human actors, while in Egypt, networks were more nonhuman, technologized and mediated. When A46 compares this with where she has settled, she describes Australia as a largely non-human network:

> I use my phone every day like say maybe a thousand times. [Laughter] I check my email every day . . . Say I want to transfer money from the bank I do it through my phone.

A44 outlines a similar trajectory, whereby her country of origin had constrained networks which limit human action:

> You can't use a public phone without a phone card . . . we only heard about mobile phone but none of my parents used the mobile phone.

This is followed by more enabling networks in their intermediate country (Egypt), where there were a range of non-human and human actors available. Her family could go to a phone shop to make a voice call or a Skype call: "The shop person would do it for me and I would talk to my father." There were also other alternatives once they were aware of what the network could offer:

> . . . then after one year, my mum, she had a home line for a few years and then afterwards, maybe for two years . . . mum had a mobile.

What is apparent in interview responses, is that networks in countries of settlement, or so-called "developed" countries such as Australia, are characterized through nonhuman actors:

> Nobody teach me, I just know it. Because everything in Australia, if you buy something there is book there, you can read so you can do by yourself. (A47)

A53 purports to feel a sense of freedom to use all technologies as much and as often as she wants in Australia, because "you are considered a human being here." That is, it is this enablement by nonhuman actors that she considers to be a distinctly human quality. Her capacity as a human actor in the network is increased by the nonhuman. This is recognized by detainees when they insist on "rights" to media. They are emphasizing their human need to access networks and actors:

> we still have rights and we want that rights even if it's a right to a newspaper or a TV or communication, some forms of communications, we should have that right and we need that right. (Issaq quoted in Fiske 2016a: 58)

The converse of this is the feeling that networks dominated by nonhuman actors alienate and dehumanize. This was apparent in respondents who were accustomed to networks consisting of primarily human actors in their countries of origin. It was manifest in difficulties becoming literate in the networks of the country of settlement, and reticence surrounding nonhuman actors, especially where technology was historically associated with surveillance of citizens. This illustrates the utopian/dystopian digital divide, whereby technology (a nonhuman actor) is seen as either liberatory or something to be feared.

It is important to remember that Actor Network Theory should not be considered in binary terms. There is no divide between actor and network, or between human and nonhuman actors. It is a model for thinking about networks as complex ecologies and topologies that are constituted by dynamic relationships between the human and nonhuman. It is complemented by Granovetter's ideas about the strength and weakness of ties within networks.

Chapter Eight

Hierarchies of Technology Literacy

The previous chapters have argued for alternatives to binary understandings of technology and users of technology. This chapter continues to build upon that theme, reframing digital or technology literacy to be more than being able to use computers and the Internet. Empirical data collected across the range of research projects discussed in this book points to a hierarchy of technology skills and literacies that refugees possess. These skills are developed (and can potentially be further developed) by refugees prior to and during displacement, as well as upon their resettlement in a new country. However, these refugee literacies tend to pertain more to telecommunications technologies, rather than computer technologies.

The key findings of the *Mind the Gap* study (Leung 2011) will be presented in this chapter. This project studied the telecommunications literacies of newly arrived migrants to Australia from refugee backgrounds. The chapter suggests that refugee skills and knowledge in relation to telecommunications technologies are also digital literacies. Though not as privileged as computer literacies, and often taken for granted, the telecommunications skills and knowledge of refugees prompts a rethinking of the notions of digital divides, in terms of highlighting how certain digital technologies are privileged over others. In light of this, the chapter finishes by interrogating commonplace understandings of digital and technology literacy.

The *Mind the Gap* research project conducted over thirty surveys and interviews with refugees who arrived in Australia between 1995 and 2009. The refugees were predominantly from the following countries and regions: Iraq, Iran, and Afghanistan; Sudan and other African countries; as well as Burma, Cambodia, and Thailand (table 8.1).

The *Mind the Gap* study had three main aims:

1. To examine refugees' knowledge of telecommunications products and services when newly arrived in Australia;
2. To investigate telecommunications literacy in refugee settlement service provision;
3. To develop a telecommunications consumer education program tailored to recent arrivals from refugee backgrounds (Leung 2011).

Table 8.1—*Mind the Gap* research study respondents: key demographic information

	Iraq, Iran, and Afghanistan	*Sudan and other African countries*	*Burma, Cambodia, and Thailand*	*Other*
Male	4	2	7	2
Female	4	7	5	1
Total	**8**	**9**	**12**	**3**
Late teens	6	1	3	0
20s	0	5	4	0
30s	0	2	4	0
40+	2	0	1	2
Unknown	0	1	0	2
Total	**8**	**9**	**12**	**3**
Arr 1995-2003	3	1	1	3
Arr 2004-2009	5	7	11	0
Unknown	0	1	-	0
Total	**8**	**9**	**12**	**3**

Leung, 2011

Refugee knowledge and skills in relation to telecommunications were the focus of the *Mind the Gap* research project because an earlier pilot study had shown that landline telephones were the most frequently used technology by refugees prior to coming to Australia, and that mobile phones (table 8.2), followed by landline telephones, were the most frequently used technologies by refugees after their arrival in Australia (Leung, Finney Lamb, and Emrys, 2009).

The *Mind the Gap* project subsequently identified a five stage model depicting a hierarchy of refugee telecommunications literacies, as outlined in figure 8.1 (Leung 2011).

In order to fully appreciate the reasons for telecommunications being the most prevalent refugee technology literacy, as well as the hierarchical nature

Table 8.2—Comparison of Australian refugee experience with various technologies: before and after arrival

Technology	*Number of responses*	*Percentage*
Landline telephone	14	33%
Mobile phone	18	42%
Letter writing	0	0%
Fax	0	0%
Email	4	9%
Internet	7	16%
Unknown	1	2%
Phone cards (could be landline or mobile)	4	9%

Leung, Finney Lamb, and Emrys, 2009

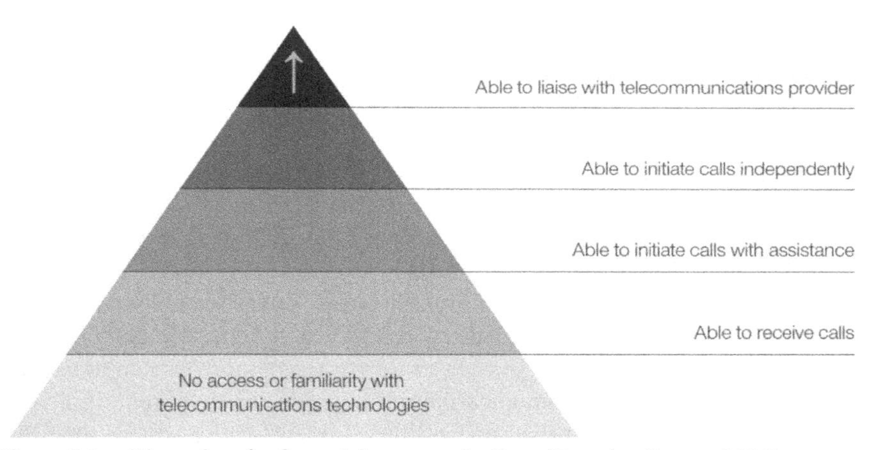

Figure 8.1. Hierarchy of refugee telecommunications literacies (Leung, 2011)

of refugee telecommunications skills and literacies acquisition, it is necessary to evaluate refugee technology experiences over three distinct periods:

1. In their home lands prior to displacement
2. In their intermediate country or countries during their displacement
3. In their resettlement country

We will now examine each of these periods in turn.

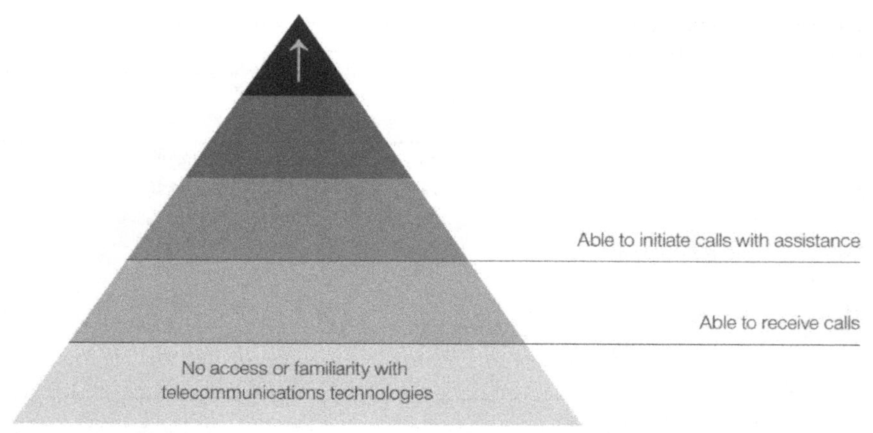

Figure 8.2. Base level telecommunications literacies with which respondents left their country of origin (Leung, 2011)

REFUGEES AND TECHNOLOGY LITERACIES PRIOR TO DISPLACEMENT

Little is known about refugee use of communication technologies in developing and/or war-torn countries of origin, prior to their displacement. However, several factors are likely to inhibit refugee access to technologies in such situations, including war zone violence and combat damaging telecommunications infrastructures, government sanctions on telecommunications during periods of conflict, and generally poor network coverage in strife-torn areas. In such circumstances, humanitarian assistance in facilitating access to communication technology is likely to reduce distress and help prevent the separation of families. Humanitarian agencies therefore have a role in the emergency delivery of letters and facilitating community access to satellite phones (Leung, Finney Lamb, and Emrys 2009).

Figure 8.2 outlines the base level literacies that survey respondents had upon leaving their country of origin. The level of literacy of refugee respondents in this study was influenced by their country of origin, with some countries (such as Iraq and Sudan) generally having better telecommunications infrastructure than others (such as Burma). At least two-thirds of the Iraqi and Sudanese respondents had acquired base level telecommunications literacies in their country of origin, compared with only a minority (less than 17 percent) of the Burmese (Leung 2011).

As can be seen in figure 8.2, the highest level attained by any refugees in the *Mind the Gap* study was the ability to initiate calls with assistance. No participants had the ability to initiate calls independently in their countries of origin, or the ability to liaise with telecommunication providers (should that opportunity have been available in the circumstances). Respondents either did not have access to any kind of telecommunications technology or access was brokered by a third party, such as vendors or humanitarian organizations (Leung, Finney Lamb, and Emrys, 2009).

According to respondents originating from Iraq in the study, prior to the wars of the 1990s and twenty-first century, Iraq had a well-developed telecommunications infrastructure. It was commonplace for respondents to have landline phones in their home. However, at times, there were restrictions on technologies for political and security reasons.

Nonetheless, some of the younger respondents recalled using mobile phones in Iraq and that Internet was also available.

> We had all the technology. We got the technology from Internet and from telephones, from computers, from like all the economy was there. (M6 arr 1995)

> No mobile. It was not permitted to anybody to have a mobile phone when Saddam was the president. (M8, arr 1999)

> Yes but not all of Iraq people had a computer at home like just like some of families. (M5, arr 2009)

> Well in Iraq we don't have already computers. (M2, arr 2009)

Respondents from Iraq did not have experience of making calls outside of the country because these were either not permitted and/or relatives would call from overseas.

There were mixed findings from the Sudanese respondents regarding telecommunications usage in Sudan. While six out of the seven surveyed said they used phones back in Sudan, both landline and mobile, more than half indicated that they stayed in touch with family members through other means, such as letter writing:

> Because we live all together we verbally communicate to tell the story. (A1 arr 2006)

> When I was in Sudan, we can write letters or visit them. (A2 arr 2004)

> No technology; we just visited each other. We would travel by transport—bus. Sometimes news would take a long time because it would rely on visits. (A3, arr 2002)

> It was a little bit difficult because we used to communicate by writing letters and then we have to possibly give it to someone to send it to communicate with the family . . . So sometime if there's anyone going who you trust to take the letter directly to your person or your family, you can send the letter easily, you could still send sort of letters in couple of days depending on who would take the letter. But you see late 90s to 2000, it become more easier because we have access to use maybe home phone especially in the Khartoum area . . . But back at Southern Sudan—maybe because there was no home phones we just use when our parents or our family members are in the office, we use the office telephone actually.
>
> Like we use our own phone in Khartoum but there (Southern Sudan) we have to make sure that it is working hours so we can communicate with them during—what do you call it—in their working hours . . . . . . From 2005 the mobiles were so spread out so everyone has mobile, it become more easier to communicate. It was anyone in all parts of Sudan. (A7 arr 2004)
>
> Landlines phones were only in offices. Some people had mobile phones. (A3, arr 2002)

As mentioned above, at least two thirds of the Sudanese respondents reported that they were somewhat literate in telecommunications technologies before leaving the country. In addition, they learned to use these technologies from various sources:

> Home phone—when the Sudatell company put it [in] they show us how to use it. Public phone—I learnt it from the company. Mobile phone—I learnt it from the shop I bought it from. (A1 arr 2006)
>
> We learnt to use these. They taught us at school or the technical people. (A2 arr 2004)

Respondents recalled being taught how to use telecommunications technologies by phone companies, retailers selling telecommunications products, and within the school environment.

Therefore, while such technologies were available, their accessibility and use were hampered by war and conflict, meaning respondents resorted to other means to make contact with family.

> I'm living in a far Western part of Southern Sudan so by that time it was closed, not any means of transport between the area because that part was closed, controlled by the rebels for example . . . because by that time (early 90s) most of the areas were closed and then we used to use the Red Cross letter . . . but that takes time to reach the destination also. (A7 arr 2004)

Respondent A7 also talks about using CB radio to communicate with colleagues who could then send an email on his behalf.

In contrast, the Cambodian, Burmese, and Thai respondents had little or no experience of these technologies prior to coming to Australia. This resulted from periods of internal displacement within Cambodia and Burma, in which communication technologies were largely unavailable and to make contact with relatives was dangerous.

> The battle between the government and the ethnicity—the ethnic group there who fight each other, and people can't stay in their [own] town. They have to flee. They move to place to place and then they lost their farm, lost their home, and then they [start] starvation and then some people become—they flee from their own village. [And then to] famine. (I12 arr 2009)

> We moved from town to town . . . . . . and that was like for three years and eight months, during the course of the war. (I1 arr 1995)

> I mean, if we are able to buy a telephone we can use but it's not easy. Even if you can, you know, because everything is controlled by government . . . if you have a telephone, and then you can accuse by the government at any time. So there's no people are not dare to buy the telephone. (I12)

This itinerant existence combined with the fear of being located meant that phoning relatives within the country of origin was avoided even if communication technologies were available and accessible. Contact was also infrequent because phone calls were expensive.

> As you know Burma is a very poor country. You can't use the telephone, Internet if you are not the member of the government. If you are not rich you can't use at all because very expensive. Only people who, they work for the government, is a member of government and if they are related to the government. (I12)

> I was really little, but they probably wrote letters to each other, 'cause the mobile phone, by what I know was only introduced more recently into the year 2000 and even then it was only the rich people that could afford the plan. (I1)

> It's hard to contact from Burma . . . you don't have [open] communication . . . when we write a letter the postmen [they were not] working well . . . We have several [times we lost] our letter. (I2 arr 2009)

> We don't have [mobile phone]. Also we don't have phone line. Just we communicate with, when we saw some people who came from our state and they will write a letter . . . Probably just one or two times every year . . . just we can tell we miss you and we want to see you, just like that. Because also we are afraid

> of the government. Yeah, we cannot tell anywhere about our situation or what's happened to us. Just only we miss you or we stay there, or something like that. (I6 arr 2008)

Similar to some of the respondents from the Sudan, the Burmese used a system of communicating with family members by having notes passed through messengers. Very urgent information could be sent via telegraph or telex, which would entail going into a post office–type establishment and paying for a message, the price depending on the number of words.

A couple of respondents claim to have never seen a telephone at all in Burma:

> Even I never seen the telephone. (I12 arr 2009)

> We never seen before . . . what you call . . . mobile phones . . . Public phones, very few times. (I11 arr 2009)

Lack of availability and security meant communicating from within Cambodia or Burma to outside of those countries was almost impossible, even if family members were just across the border. Stories of families having no contact for years were commonplace, even if the distance was not great.

> We can't because they didn't have a phone also and we never contact. We never communicated. (I4 arr 2007)

There were also instances of borrowing technologies from those who had access to it.

> In our village, we didn't have anything—just only camera really . . . Not ours, just my friend . . . we didn't have any electricity as well. (I5 arr 2008)

> We don't have television. Just one or two—for example in village because I live in small village, just two or three people have television and then we have to go to their house and then watch it. (I6 arr 2008)

> We had to go and borrow—like every time my uncle called (from Australia), he would "prank" a call to the neighbours (arrange with the neighbour fetch the family to receive a call at a later time). (I1)

REFUGEES AND TECHNOLOGY LITERACIES DURING DISPLACEMENT

Once refugees leave their countries of origin, telecommunications literacy acquisition during displacement was significantly impacted by both the intermediate countries where they were located, as well as the circumstances of that displacement. For example, the majority of Burmese refugees in the *Mind the Gap* study were displaced to refugee camps in neighboring Thailand, with little or no technology for communicating with the outside world. By contrast, refugees in that study from Iraq and Sudan who had spent their displacement time in intermediate countries such as Jordan and Egypt were generally exposed to new technologies, including telecommunications. Significantly, the study found that those who had spent as little as one year displaced in a more conducive location may acquire an additional telecommunications literacy, such as the ability to initiate calls independently (see figure 3), compared to refugees who had spent up to twenty years in refugee camps, where telecommunications technology literacies tended to stagnate at the lower levels of the hierarchy (Leung 2011).

Respondents in the *Mind the Gap* study originally from Iraq fled because of the wars in their country and were displaced to intermediate countries before arriving in Australia.

> We have to leave Iraq to Syria. My neighbours help us help my family and me and we go to Syria. We live in Syria three years. (M2 arr 2009)

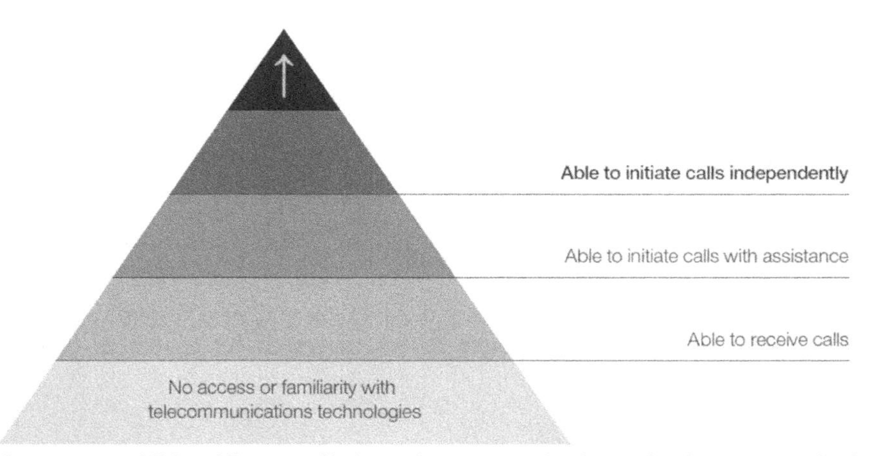

Figure 8.3. Additional literacy of independent communication technology use acquired by some respondents while in their intermediate country (Leung, 2011)

> I left my country in 2003, I went to Syria for one year then I came back to Iraq then I stayed in Iraq about one year and then I went to Jordan. I lived in Jordan about five years or six years then I came to Australia. (M5 arr 2009)

Respondents from the Sudan left because of the war and political persecution and most, but not all, came through Egypt before coming to Australia:

> For me I can't come to Australia direct because we don't have an Australian embassy in Sudan but we have British embassy it can help us in case. From Sudan I flight to Cairo and then I spend like five months, two weeks in Cairo and I went to Egyptian embassy to make my process to come here. (A1 arr 2006)

> I come from the Khartoum city of Sudan to the Halfa border of Egypt by train. Then I caught [a] ship from Halfa to Aswan. Aswan is a port of Egypt. Then I caught a train from Aswan to Cairo. Then I met my husband at train station. I almost took eight months in Egypt after that. My husband and I travelled to Australia by plane from Egypt to Dubai.

> From Dubai to Manila. We stay six hours in airport. Then we travelled from Manila to Sydney. (A4 arr 2004)

> I went from Sudan to Egypt by boat. I stayed in Egypt for two years and then came and settled in Australia. (A5 arr 2005)

> From Sudan I went to Cairo. I was [in] Cairo three and a half years. Then I came to Sydney. (A6 arr 2005)

> We leave Sudan because my husband use to work as a journalist and the government put him in the prison. When he come out he decided to leave Sudan to any country. We went to Lebanon and then to Australia. (A2 arr 2004)

> Because the conflict in my country was mainly between south and north . . . we Southerners who move from our area to other area considered us IDPs, internally displaced people because we're not in our area . . . I took a refuge to Egypt . . . I stayed in Egypt for nearly one year and a half before moving to Australia. (A7 arr 2004)

According to Cambodian and Burmese respondents, after being displaced within their countries of origin, their families escaped into neighboring countries, principally Thailand, to live in refugee camps for a number of years before migrating to Australia.

> Yeah, so it was like four years of lost contact. 'Cause when he lived in the refugee camp, we didn't really know much about him. There wasn't any communication there. (I1 arr 1995)

> In 1990 we had to flee into Thailand and then we stayed in Thailand around 18 years in a refugee camp. (I6 arr 2008)

> Also my grandpa was like . . . a soldier . . . so he can't go back in the Burma country . . . when he was going in there some people will kill him or something. So he didn't go in so he moved around and they moved to Thailand. (I5 arr 2008)

Because the various countries to which the respondents fled were so diverse, experiences of technology in those countries was similarly very wide-ranging.

In Jordan, M5 recalls having access to mobile phones, phone cards, and Internet and being able to easily stay in touch with relatives in Iraq, Syria, and Lebanon (but not landline, as they did not have their own).

M6 also reported using telephones and phone cards regularly in Jordan, as well as Bulgaria and Greece, as well as faxes.

Like M5 in Jordan, M7 says that while in Syria they did not have their own landline, so would use mobile phones to contact other displaced family members, or would go to a shop where you could make telephone calls.

M2 says that while she was in Syria all contact with family abroad was conducted through the landline telephone and phone cards, although this was infrequent, and involved receiving calls only. Her family did not have a mobile phone.

M1 remembers having a home phone in Lebanon, and that her family used phone cards. The Internet was also used but only in establishments similar to Internet cafes or shops. Her father had a mobile phone. Like M6, she says that faxing was also more common than letter writing, which was practiced as well.

Respondents originating from Sudan who had been displaced to Egypt clearly had more consistent exposure to telecommunications technologies in this intermediate country. All stated they had experience of using landline or home phones, public phones, mobile phones, and phone cards in Egypt. Respondents had to either teach themselves or be coached by family members in learning to use these technologies.

A3 recalls using a phone and phone card for the first time in Egypt: "I first use the phone in Egypt to call my sister with the phone card. The phone card was easy to use. I followed the prompts in Arabic."

> I just lucky to also—to get—what do you call it—all the means for the new technology handy. So we have telephone at home—home phone—and then the other

> thing is I was lucky to get some employment, to work as an IT—what do you call it—as an office assistant. So basically I was working on the computer with all the Internet facility for the first time. So maybe in Egypt this is where I establish my first email address here. That's pretty cool, hey. Then also in Egypt, I start using emails as a means of communication, we start using—instead of telephone we start using—what do you call it—I don't know what you call it—the phone which is connected through the internet. (A7 arr 2004)

> Yes, we always had a phone (in Egypt)—a home phone—but my parents they didn't have mobile phones. I know how to use home phone. (A9 arr 2005)

Because telecommunications facilities were more accessible and affordable than in their country of origin, Sudanese respondents who had been displaced to Egypt found themselves having to bear the cost and responsibility of staying in touch with family members in their country of origin:

> We'll like [buy] the card from Egypt when we'd call them . . . I would call them—I'll like set up on that day we're going to call. So like all of them they come to one house, one of my cousin's houses so we can talk to all of them. (A9 arr 2005)

The Thai refugee camps had little or no facilities for communicating with the outside world, as there were no phones or computers whatsoever available.

> In the city, yeah, they use mobile phone. But in our village, no. No connection as well I have never see a computer when I was in Burma; never, ever, seen computer. (I5 arr 2008)

> In Thailand because we live in the refugee camp so we don't know how to use the phone. (I4 arr 2007)

Those with relatives who had emigrated to countries of settlement occasionally had mobile phones, although this is described as "very very rare" by I10 (arr 2009). Refugees who did not have their own phone could pay to receive calls on other people's mobiles. Calls were generally initiated from the outside as making calls was more expensive than receiving them.

> When I lived in Thailand I didn't have mobile . . . Just like a big (public) telephone. (I5 arr 2008)

> They contact us, yeah . . . Because in camp we didn't have enough money to contact them . . . Yeah, it was too expensive to contact in the other country. (I4 arr 2007)

As calls were mainly received, respondents had little experience of making calls. Furthermore, on rare or emergency occasions when calls were made, it was done in a shop/call center environment where the making of calls was assisted or done for them:

> When we use, the owner press for us, just [we show] the numbers and they press for us . . . operate everything. (I6 arr 2008)

In certain circumstances, respondents worked for NGOs or would go to school outside their refugee camps. This exposed them to technologies such as computers and mobiles phones, with which they were free to use and experiment.

> When I finished high school, I moved to—because I wanted to improve my study as well because I move to another camp to study there. That have connection phone . . . Because of my teacher that I live with him, because I always help him in his house so he give me one (mobile phone)... I just copied him and looked at what he did and how he did it. (I3 arr 2007)

> Because I work with NGO, yes, they showed me a little bit. [Then I play around] here, I know how to use. (I10 arr 2009)

> Later on around 2005 I am working outside the camp . . . I can call coordinator—he provided a mobile phone for me because we worked together and then he provide for me. Then he showed me how to use it . . . I learn Microsoft Office, I just have to learn a little bit, not too much. (I6 arr 2008)

For most respondents (those originating from Iraq or Sudan), intermediate countries exposed them to new technologies as well as new skills and literacies in using them. This was not the case for those who had spent significant periods in refugee camps (those originating from Burma). Therefore, there was no correlation between amount of time spent in intermediate countries and level of technology literacies. Those who spent as little as one year displaced might have acquired the additional literacy of being able to initiate calls independently while those who had spent twenty years in a refugee camp found that their technology skills stagnated at the lower levels of the hierarchy.

REFUGEES AND TECHNOLOGY LITERACIES AFTER SETTLEMENT

The *Mind the Gap* research project identified refugees in Australia as vulnerable telecommunications consumers (Leung 2011). As new arrivals to the country, being at the lower end of the telecommunications skills hierarchy places them at a distinct disadvantage on at least two levels. Firstly, living in contemporary Australia demands higher level literacies (such as owning and taking financial responsibility for a range of telecommunications and other technology products). Secondly, given the importance to refugees of keeping in touch with displaced family and friends, telecommunications technology knowledge and skill development is vital for their emotional well-being.

In terms of living in contemporary Australia, the Australian Communications and Media Authority (ACMA 2015) reported that digital communications are now embedded across Australian business, work and social life, with 92 percent of Australians having used the Internet in the six months to May 2014, including 100 percent of those aged eighteen to forty-four. Significantly, the study further found that the way Australians access the Internet is changing, with 70 percent going online via their mobile phone. This latter finding suggests that refugees' inclination toward telecommunications as a communication device could also potentially facilitate their greater use of the Internet, participation in society, and employability, via that medium. The capacity to "catch up" with the majority of the Australian population's technology use is further enhanced by the 70 percent of refugee and humanitarian arrivals being aged under thirty years (Migliorino 2011).

While many refugees arrive in Australia with a base level of telecommunications' consumption literacies (that is, the ability to receive a call), there was no subsequent formalized training or support for the further development of their telecommunications knowledge, skills, and service provision during the settlement process. The refugee settlement process had previously focused on the development of computer literacy and English language skills, not telecommunications. However, the *Mind the Gap* study concluded that computers are not the most appropriate starting point for increasing the technology literacy of refugees, as telephones (home phones, mobile phones, and phone cards) are generally seen as more resonant and meaningful technology tools for staying connected with loved ones overseas. It was recommended that refugee learning and development of communications technology literacies be formalized as a part of immigration policy and the settlement process (Leung 2011).

All respondents in the *Mind the Gap* research project came to Australia with family members, and so, had networks of support for learning communication technology.

Landlines

Respondents described their first experience of having and using a home phone when in Australia:

> It's very hard to do it in my first home. But, yeah, somebody will come and show us how to do that and they're like you might actually want to contact with friend in Thailand.
>
> So we want to do that but we don't know how to call them. We don't know how to use the [country code] or something like that. So the other friend, they told us you have to buy full coverage like that. Then you have to use this [country] code to call and then press this key with that key. (I3 arr 2007)

> Like because when we live in Thailand we never have a home phone. Yeah, just here and then like when we heard the phone ring we are looking at it [laughs]. (I5 arr 2008)

Since arriving in Australia, all respondents had used landlines in the form of home phones, despite some never having a telephone in their homes prior to coming to Australia. As part of the initial settlement process, it appears that respondents were mostly set up with a home phone in their new accommodation. The home phone was a more foreign technology than mobile phones for a number of respondents. This was a factor in some electing not to have home phones after a period of time in Australia.

> It cost me a lot of money (around $40 per month) . . . for me very expensive because I have to pay my electric, gas, water and telephone, house telephone and mobile. So each month I have to pay a lot and cost is really we don't need . . . So right now it's cost me a lot of money (last month $60) so I decided not to have any more. But I have the mobile phone. (I10 arr 2009)

In the Australian context, it was clear that participants struggled to understand their phone bills, how costs were incurred and the contracts they had with their telecommunications provider.

> Now we are dealing with Optus, but before with Telstra . . . they bill us for telephone number which we never use it . . . How can we use—who I know in Afghanistan? I don't know nobody there . . . I wouldn't even know how to call Afghanistan (laughter). (M8 arr 1999)

> Before we had Telstra but the speed wasn't very good. Then we changed it to Optus. Now it's good... Yeah just for speed and sometimes when we speak with our family their voices wasn't too clear. My father said maybe it's poor line, so we changed it. (M4 arr 2002)

> First I arrived, my home phone is Telstra . . . My uncle changed it and [because] if I do Internet, Internet bill was $75 and for phone bill $80. (I7 arr 2008)

One respondent, O3, in attempting to have his home phone connected with Telstra, spent so long waiting on his mobile phone trying to get through that he used his entire $30 of prepaid credit. After that experience, he decided against having a home phone. He also disliked the requirement of telecommunications providers to directly debit his bank account.

Mobile phones

Generally, the Iraqi respondents had previous experience of using as well as owning their own mobile phones prior to coming to Australia. For respondents from the Sudan and Burma, Australia is the first place where they have *owned* their mobiles, although they may have used them in the past. Respondents waited anywhere between two weeks and three years before buying their own mobile phones. In some cases, settlement organizations equipped families with a mobile phone where there was no landline in the temporary accommodation:

> When you come to Australia there is people that help you and they gave us a mobile phone to use it for emergency. (M1 arr 2009)

> When I come arrive I be give new one—very old one—mobile phone . . . So can ring case worker all the time . . . [After a while I] buy for myself and they will give back to worker. (I10 arr 2009)

> They gave us one only to contact with case worker can call us or we can call to our case worker. Yes just only in . . .for the group . . .in the community. (I11 arr 2009)

A settlement worker for older refugees observed that his clients were limited users of mobiles, using them only for voice calls and rarely did they know how to use SMS.

All respondents had acquired and were continuing to use mobile phones since arriving in Australia.

> When we first came to here, then one of our friends—because we don't know how to buy the phone or nothing like that—so he bought for us. Because I don't know how to use the other mobile phone in Australia because it's really hard to do for me personally. So I just have my own bought for me, the Nokia one because I know how to use that. (I3 arr 2007)

Respondents who previously had mobile phones before coming to Australia, also bought new ones after arrival but would also use handsets purchased in other countries by inserting a new SIM card.

> I had one from Jordan but I just change it when I can. (M5 arr 2009)

Experience of using mobiles prior to coming to Australia did not appear to inform the learning process. Overwhelmingly, respondents were taught how to use mobile phones informally by observing friends and relatives or having them actively demonstrate functionality.

> I just copied him and looked at what he did and how he did it. Then, yeah, I tried to do that. (I3 arr 2007)

In rare cases, respondents who had no previous experience of mobile phones taught themselves through experimentation, trial and error: however, this was extremely uncommon, and confined to young people from refugee backgrounds.

> Honestly I learn by myself. (I12 on texting)

> At first it was hard like to go to the settings and tools but now I tried by myself to go there and I learned how to use them. (M1 arr 2009)

All mobile phones were individual objects that did not have to be shared with others: each member of the family had their own handset. A few respondents had more than one mobile phone each. One respondent carried at least two mobiles with different providers so that he is able to get coverage no matter where he may be. Another did this to take advantage of free talk time with users on the same network. For example, I2 had one mobile with Vodafone, the other with Optus. The mobile was regarded as the key technology for staying connected with immediate family members with them in Australia, so that they are accessible, particularly in any emergencies:

> I used mobile phone because everyone want [to] call. They will find me anywhere. (A1 arr 2006)

> I use mobile phone to contact my friends and my husband if he is out . . . if I need something from the shop or if I want to go somewhere. (A2 arr 2004)

> Sometimes [when] I away I use it. (A4 arr 2004)

> My husband just bought it for me because if I needed help or to know where he is and I want him to buy me something from the shop. (A3 arr 2002)

There were instances of respondents not wanting a mobile phone, as they did not see a need (O1, I1). However, family members insisted they needed one so that they could be contactable:

> When I went shopping for buy a mobile phone, I didn't like all the phones. They were all ugly . . . Finally, the last time when I went to the shop for buy a mobile phone my mother said I don't care, you have to buy one. (M3 arr 2009)

> My father said it's a very important thing so you have to do it, to buy. (M4 arr 2002)

> Because if we—sometimes if we are walking we need a mobile. If my aunty calls . . . don't know where we are. (I4 arr 2007)

> Because most of my family members, they have their own phone and then sometimes I can use their phone so I thought it's not necessary for me yet . . . and I see that, you know, in here. This is—sometimes, you know, I need for myself, like when I go out and I probably have some problem or . . . and then sometimes, you know, because . . . privacy. (I12 arr 2009)

Mobile phones were also used for making international calls to family members overseas in urgent circumstances.

> Whenever it's necessary because it's not cheap. (I12 arr 2009)

> Just once before when I [. . . in 2008] had one of my brother pass away in my country so I had to talk . . . on my mobile. (I2 arr 2009)

In other cases, the mobile phones were used with particular phone cards like the Optus $10 international calling card:

> Yes, that's for 200 minutes for AUD$10. So Optus pre-paid is very useful for international calls to Malaysia, Thailand. (I11 arr 2009)

> Ten dollars you can call three hours. Yes, so very useful . . . China, Canada, US or Thailand, Hong Kong, Malaysia, Singapore. (I10 arr 2009)

In cases where respondents had more than one mobile phone at a time, one of the phones was used exclusively for making international calls, usually with a Lebara prepaid SIM card:

> We have two mobile like one for the Optus and one for Lebara. (A9 arr 2005)

Mobile phones were primarily used for voice calls. The data indicates that there was wide use of bluetooth for exchanging music files, and that listening to songs was the second most used function behind voice calls.

SMS texting was used to a lesser extent by a third of the respondents. This was not necessarily age-related, as some younger respondents also reported that they did not regularly send text messages. It is possible that this is due to low levels of English literacy, rather than a lack of technical ability. Therefore, it could be surmised that there is a preference for functionality that utilizes aural literacies (such as voice calls and music) over reading and written literacies (such as texting and Internet searches).

For those that did not use the Internet, there was a perception that accessing the Internet on a mobile phone was unnecessary (particularly if it was available on the home computer), and furthermore, that it was relatively more expensive.

> I don't use the Internet on my mobile phone, 'cause I don't feel the need to. 'Cause I can access the Internet at home and when I'm at work or when I'm at school. There's the Internet everywhere, I don't want to waste money with the net connection. What else? To listen to music—I have a memory card and store music there. (I1 arr 1995)

> Because I don't know how to use Internet on phone and I don't know—because some people they know how to—how much they have to pay when they use their mobile phone for Internet. But for me no, I didn't know yet so I don't want to use it yet. If I know more information I will use it. Because sometimes it's very dangerous for us when we don't know the information right then there will be big problems. (I6 arr 2008)

> Oh no, no. I never use the Internet—Internet on phones . . . I know that this quite expensive to use the Internet on the phone . . . I didn't press and they say the Internet automatically come out and then all your money is gone after a minute. So I was so surprised. Yeah, it's a very expensive [mistake]. I [didn't press this] on purpose, you know, some kind of [accident] and I see that and then I—the—finally when I checked the money [because the credit is gone]. (I12 arr 2009)

A2 says that she does not use MMS, email, or Internet from her mobile phone "because they are expensive." Where Internet is accessed from a mobile phone, this was done by the younger respondents on prepaid plans. Seven young people (M5, M1, I5, I3, M4, A9, and M3) from refugee backgrounds indicated they used mobile Internet mainly for Facebook (including sending email) and YouTube. These respondents were also attending school and heavy users of the Internet at home and school).

Phone cards

The vast majority of respondents used phone cards. Those that initially said they did not use phone cards, actually used discount SIM cards in their mobile phones to make cheap international calls. While phone cards can be used with both landlines and mobile phones, SIM cards can only be used in mobiles. Phone cards and SIM cards are technically different, but are adopted by respondents as cost-efficient alternatives for making international calls. It was more common for respondents to use phone cards with landlines than with mobiles.

While some respondents had used phone cards in intermediate countries (A7, M1, M2), respondents more commonly only discovered and used phone cards after arriving in Australia, having been recommended them by word-of-mouth through their communities.

> The corner shop guy is a friend obviously, because we always go there and buy a lot of stuff. And he's like oh try this phone card because it's much cheaper. That was the first time they introduced the phone card. It was more reliable and now that there's competition within the phone card, once they establish customer and whatever, actually the service decrease and then she will find another phone card that's more competitive. Then it just moves in a cycle. (I1 arr 1995)

> My father have friend here and he showed my father to buy . . . and then how to use that. He teach all my father . . . and then my father learn for to show my sister; my big sister and then my big sister show me. (I5 arr 2008)

Respondents told of cousins, aunts, uncles, mothers, fathers, sisters, brother-in-laws, family friends, neighbors, caseworkers, and interpreters recommending the use of phone cards. In many instances, they learned through relatives who had been calling them using phone cards.

> I was in Burma because every time the [relative call] they using cards. (I2 arr 2009)

> My father knows about that because always he used to call us . . . his friends told him if he want to call overseas to your family, so you use this phone cards, that's easier to use. (M4 arr 2002)

Respondent I12 talks about being shown how to use the phone card upon arrival in Australia, but he had not even used a phone yet. That is, he was taught how to use the phone card having never used a home phone previously. In one case (I3), the home phone was restricted to local and national calls only, therefore, the only option for making international calls was to use a phone card.

While a settlement worker commented that phone cards were very "multicultural" in that they catered for different languages, there was evidence to suggest that respondents needed assistance in using them at least the first few times.

> In the back of all the phone cards they have how you can use it. But when we first came to Australia, we hadn't enough knowledge about how we can write in English. But my father learned me how to use that. (M4 arr 2002)

> I go to my family friend to do it for me . . . Very hard but now I'm used to it, they teach me . . . I never knew how to use it so they show me how to use it . . . because my Dad he's the one used to calls so when I buy I don't know how to use it. I went to my family friend house so they used to do it for me. (A9 arr 2005)

> When we got the card, we had to follow the instructions. Sometimes I was bad with my English . . . Sometimes we have to press the PIN number. We had to enter the PIN number and if we did the wrong number, it's not there . . . When I would recharge it, oh, I was so confused with it. (I3 arr 2007)

> The first time is really hard because I not like how we have to do. Is it like number one, two, three or a lot the number . . . and then you have to read. Because the first time I can't read English properly, yes, then just guess and then ask my sister. So she showed me. (I5 arr 2008)

> Actually it was my sister, my older sister, she showed me how to use that at first, but and after that, but you know, it's not for me. It's not easy for me to remember, you know, and I have to try it by myself several times and then, yeah, it take about a couple of weeks . . . I am calling to one of my friends to Thailand. First I have to call the local number first and then I call the—put in the pin number and type the number what I'm calling to and then, yeah, it's hard, and also the code. (I12 arr 2009)

> You only like spend the first two weeks are hard because you don't know how to use them. But then, when you practice it's easy. (M1 arr 2009)

> The first time yeah, difficult. Because we had to dial the city code, the state code . . . and then we have to insert the card number and then later on we have to insert the number that we want to call . . . So probably three steps so I have to write down the steps. (I6 arr 2008)

> At first complicated for me. When you listen to the [unclear] instructions it's very clear for me. But if you don't understand instructions, if you can't read these things then very difficult for the other people. So we have to approach a person who understands very well. (I10 arr 2009)

Respondent I10 goes on to explain that phone cards are so difficult for some members of his family that they have to use the more expensive but simpler option of international direct dialling, when no one is available to help use a phone card. For other respondents, there were ongoing difficulties with using phone cards:

> I have to use for my grandparents. They can't read the tiny print on the card. They can't read it . . . So I have to call for them. (I1 arr 1995)

> Yes you can find Spanish, Arabic, China, Indonesian language, any language, but not Persian. (M3 arr 2009)
>
> Yes, difficult . . . I do not understand English . . . I have to listen to English, but too hard. (I7 arr 2008)

In addition to the problem of understanding the verbal instructions in English, in one respondent's (I12) case, there was also fear that doing something incorrect would have dire consequences:

> Sometimes I feel like . . . if I press and then . . . something happen. You know, I suppose they will hear and if we just press something wrong and it—we must have that, international police or some police or some [emergency], they will come and they see you. (I12 arr 2009)

There were cases of loyalty to a brand, where respondents repeatedly bought the same phone card. However, brands were not necessarily recognized by name, but rather by how the phone card looked:

> Olive and—I don't know, it's red. I don't know if you have seen that before . . . with fire. (M4 arr 2002)

> Plus the most good is a green card that you can use for one hour . . . for Burma (I11 arr 2009)

> They didn't use the same one like when the first time they used, I didn't remember the name. The symbol was with the chicken one but now they use new—their card is like blue. (I3 arr 2007)

Where there was no loyalty to particular brands of phone cards, it was largely because respondents had repeated experience of them being unreliable. However, the potential cost savings are worth the risk of the relatively small investment of AUD$10 in a phone card:

> She's (mother) probably loyal to one for a couple of months until she's like, oh you know, when you get disconnected and then they're cut a lot. They're like

you have 200 minutes and then you get disconnected within the first 15 minutes instead of saying you have 185 left. Oh you only have 140 minutes remaining. And she's like what? So yeah, they're a bit smart now with the phone card. (I1 arr 1995)

When you use this 60 (minute card), it'll come to 40. Most of this is about 40 minutes, 30 or 28 minutes, so I can change up because it had 67 minutes, like what they told me in Kenya. When I come to use them, my own phone did just, cut off at 26. (A19 arr 2006)

It depends whether you call which company, you buy, its always got it's own problems. They cut off in the middle of the conversation and some of them have got a connection fee and probably that's why they cut off and when you ring again, whatever, its all different, but it is still cheaper for me. (M9 arr 1984)

If you buying card and then if you like to call them . . . that finish quick. If you try with . . . another one, they'll be better than the other ones. (A22 arr 1998)

Other card, you recharge it today you talk like 30 minutes, after a minute, tomorrow you didn't get it . . . Some card you buy it, when you need to recharge it, it didn't get any of the money inside. You need to call back and talk to the company, why I buy the card and didn't get anything? (A24 arr 2004)

Sometimes you press the number more than once. On the card, it said number already been used. (A44 arr 2006)

Any one I buy if it does what I want it to do then I'll buy it. If it doesn't do the job I want it to do, I don't buy it . . . If it doesn't work I change it . . . Sometimes in Call Mama just they hang, just after one minute it finished. (A35 arr 2010)

Still you're dialling, dialling and then disconnected, disconnected, they say to you, you have to ring to customer service and then before you use your money gone, you didn't connect. (A37 arr 2003)

Respondent O3 also found that after using a phone card after some time, the quality of connection degrades, so he will then move to another brand. Because of the unreliability of phone cards in terms of quality of connection and amount of talk time, word-of-mouth recommendations were important:

I ask like how many minutes for overseas to Jordan or Syria and they say they like this one better than this. (M5 arr 2009)

People who have been using it tell you, this card is good. This one it's not good. This one you have more credit on it. This one, no. This one is good for this country. This one is good for that country. (A28 arr 2003)

The number of minutes the phone cards claim to have for AUD$10 was an incentive to purchase. Phone cards that claim to have more minutes for the same amount of money was a motivation to change brands.

> I can call to Burma AUD$10 for half an hour but the other card we can use only 15 to 20 minutes. (I2 arr 2009)

> My grandma . . . she would shop around the whole Cabramatta just to get a 50 cents cheaper phone card. (I1 arr 1995)

An alternative to phone cards used with the home phone were prepaid SIM cards. Respondents discovered that the latter could be an even cheaper means of making overseas calls. However, as this required some knowledge of removing and inserting SIM cards into mobile phones, this was not as prevalent among the respondents although some did have two mobiles, one of which had a prepaid SIM such as Lebara installed to be used only for international calls. Where A9 did not know how to use phone cards as her father was the one using them, she nevertheless found a SIM card relatively intuitive to install and use:

> Easy, I get to read and I did it. (A9 arr 2005 on using a Lebara phone card)

Online platforms

Respondents were open to using new technologies, but cost savings were weighed against whether it was a practical option for staying connected with family members abroad. That is, they were constrained by the accessibility of these new technologies to their relatives overseas, which was further limited if those relatives remained displaced or in refugee camps.

Almost all respondents were users of the Internet, to which they were mostly introduced only after arriving in Australia. Uses of the Internet included email, instant messaging, and VOIP calls. However, these were generally not technologies used to sustain connections with familial networks abroad. For example, while the Internet may be cheaper still than using phone cards, it was impractical to keep contact with family members abroad as access to the Internet at the other end was problematic:

> With the mobile phone they can call and they can be anywhere. For Skype they have to be in front of a computer. So it's very inconvenient. But most of the time we use phone cards. Rarely we use Skype even though it's free and it's better, but yeah rarely . . . the thing with Skype is the other people that we're communicating have to have a Skype account and not many of the relatives who live in

> remote villages and stuff—they all have mobiles, they don't have a computer, the Internet—'cause it cost a lot to join the Internet in Cambodia. (I1 arr 1995)

While younger people from refugee backgrounds were more comfortable with using email and Internet to stay connected with friends and family around the world, their parents more regularly relied on landlines. Younger respondents were also exposed to daily computer use through school. Although young refugees would maintain connections with members of their family abroad, this tended to be with relatives of approximately the same age and of the same generation. Furthermore, they did not assume the primary responsibility for sustaining relationships within family networks; instead, the parents did so with the most appropriate technologies that were available and affordable, shouldering the financial and legal liability for each.

> My parents don't email [relatives]. I email my cousins and stuff, back and forth, but my parents no. They're not very technical with the keyboard and stuff. (I1 arr 1995)

> My daughter use it but I don't know—she talks with her friends and sometimes with a webcam. It's called webcam? (I2 arr 2009)

Where technologies such as Skype, chat, or email were used, younger members of the family had to help facilitate this communication for their parents. In addition, the family members with whom they were communicating had already migrated to city areas or countries of settlement:

> Just in the Thailand, just in the city and some camp they have computer. So they can talk with us but not really to see face-to-face . . . they do have Skype but they have to pay lots of money for that if they use the Internet. So we usually email and send pictures. (I3 arr 2007)

TECHNOLOGY LITERACY VS. DIGITAL LITERACY

The types of telecommunications literacies that refugees possess have tended to be overlooked in existing digital literacy research literature. Digital literacy has been generally narrowly defined as a person's ability to use a computer and keyboard, or how to do online searches (Gurak 2001), rather than other technologies. Livingstone et al. (2005) describes a digitally literate individual as one who can search efficiently, compare a range of sources, sorting authoritative from non-authoritative, and relevant from irrelevant. Buckingham (2007) contends that such definitions are limited, arguing that digital or computer literacy is often measured in terms of a minimal set of low-level functional skills that enable a person to operate effectively using software tools, often in performing basic information-retrieval tasks. In addition to pointing out the need for a broader reconceptualization of the notion of digital literacy, Buckingham (2007) also argued that information and communications technologies (ICTs) are no longer confined to desktop computers, or necessarily computers at all.

Indeed, the ways in which digital literacy have been defined sets up the kind of social exclusion it purports to address. Yet digital literacy is now regarded as part and parcel of being an active citizen in contemporary society (Kluzer and Rissola 2009; Migliorino 2011). Because refugees' technology literacies are not recognized as legitimate, they are inevitably positioned on the "wrong" side of the "digital divide."

In summary, this chapter has shown that a refugee's country of origin informs their technology literacies. Displacement can increase these technology literacies if refugees are exposed to new technologies, but for those who had spent lengthy periods in refugee camps, this was not the case. Therefore, refugees arrived in Australia with base level literacies where successful settlement demanded higher level literacies that involved managing complex transactions and interactions with service providers that included setting up accounts for home phone, mobile, and Internet access, as well as managing spending on communication technologies. That is, both financial and technological literacies are required at the higher end of the hierarchy.

Refugees were used to going without technology (when there is none available and/or they do not have the finances to access it) but have also experienced fear of technology and can associate it with danger. However, many were compelled to use technology nonetheless, especially as settled refugees usually bear the cost and responsibility for staying in touch with displaced family members. The majority of respondents used home phone and phone cards to maintain connections with family members abroad, while mobiles were used to stay networked with family in Australia. Respondents favored functionality which utilized aural literacies (voice calls, music) over written literacies (texting and Internet use). Refugees were not heavy users of Internet when it came to sustaining familial networks. Such comprehensive insights into the technology literacies of refugees should enable the design of services that are inclusive and appropriately targeted.

Part IV of the book explores the principles and practices of human-centered design, as well as specific examples of grassroots technology projects in which innovative solutions have been developed and engineered by refugees themselves.

Part IV

PRACTICES AND PRINCIPLES

Where part III offers nonbinary theories as alternatives to dualistic models traditionally used for thinking about technology and its users, part IV presents complementary principles-based practices that can be used on the ground by those working with refugees. Chapter 9 examines the principle of access as multidimensional and distinct from the notion of availability. Chapter 10 explores the concepts of user-centeredness and usability. This is followed by discussion of examples of user-centered design in action, and the methods and tools deployed.

Chapter Nine

Accessibility: Moving Beyond the Disability Paradigm

Previous chapters have discussed the importance of providing technology services and platforms that are familiar to refugees. This approach advocates removing constraints that impede refugee access to technology services, as well as to their subsequent use of these services. This chapter discusses the concept of accessibility more specifically and argues for a broader application of the term.

Traditionally, accessibility has largely pertained to making the World Wide Web accessible to people with disabilities. However, it has wider relevance to other groups including refugees, as accessibility is conceptually underpinned by an ethic of inclusion that directly contrasts with the exclusion inherent in "digital divides." Those concerned with providing services that are truly accessible and meaningful to refugees need to look beyond technology disciplines. In particular, design disciplines offer ways of thinking, and approaches to experience design and service design that are explicitly user-centered and customer-focused, which could potentially be applied to interventions that improve the accessibility of technologies and services for refugees.

This chapter looks at the principles and practices of enhancing the accessibility of technologies to refugees. It contends that any kind of technology access must consider the capacities and literacies of the end user. Therefore, in relation to refugees, accessibility has as much to do with the language, financial and technical literacies of users, as the technologies that are available to them.

DISTINGUISHING BETWEEN AVAILABILITY AND ACCESSIBILITY

Availability is about the supply or provision of a technology. *Accessibility* refers to the extent to which available technologies can be accessed (Leung 2014). The following quote from a respondent illustrates the lack of availability of ICTs in their country of origin:

> No, never. No mobiles. No landline phones. (A54)

From A54's recollection, there was little to no communication outside of her town or country. By contrast, lack of accessibility is exemplified in the experience of A53, who says that she couldn't call her family directly, as they didn't have a landline for security reasons. Instead, she would call her family's neighbors who would then go and look for her father or other family members. Often she would have to wait 10–15 minutes to talk to someone. On a few occasions, the Arab families for whom A53 was working as a cleaner, allowed her to use their landlines to call her family back in Somalia. Accessibility is constrained at both ends: there are landline telephones available to A53's family, but they fear the consequences of its use; while for A53 herself, her access to a telephone is brokered through her employer.

Thus, simply making a technology available does not make it accessible:

> Unlike roads, the provision of digital connectivity is not sufficient to ensure the empowerment or even equitable inclusion of the target population. (Prasad 2013: 229)

Warschauer (2003) argued that there are two main models for understanding technology access. The first is in terms of the simple physical presence of a device. The second is in relation to conduits: things that facilitate use of a device, that is, a network of some kind such as electricity, Internet, or telecommunications infrastructure. Both these models are arguably more about making technologies available than accessible.

Warschauer (ibid.) proposed a third model for understanding access: literacy, defined largely in terms of language (reading and writing) and technology skills (digital literacies). This model can be differentiated from the previous two in terms of its concern with accessibility as it is discussed in this chapter and the role of literacies. In addition, financial literacy—which is not referred to by Warschauer—also plays a significant role in technology access (Leung 2014).

In distinguishing between availability and accessibility, it is then possible to understand that the ratio of one to the other can be configured differently.

For example, there may be widespread availability of a particular technology, but the capacity to access it can be constrained by factors such as:

- the affordability of that technology
- government restrictions on use of a technology
- levels of literacy in that technology

Principles of accessibility have predominantly been associated with designing products or services which can be accessed by all, regardless of ability, and in ways which take into account diverse use (not just conventional use). Part of this ethos of inclusiveness—designing for access to all—means considering the needs of a range of so-called "minority" user groups and the commonalities between them. As such, applying principles of accessibility goes beyond issues of disability, the sector where much of the debate and discussion about accessibility has taken place.

Looking across the literature on technology and inequality, which examines the use of technology by minority groups, it becomes clear that refugees are one of a number groups and communities that are consistently positioned on the "wrong" wide of the "digital divide." Aside from new migrants and refugees, they include people from non-English speaking backgrounds (NESB) and culturally and linguistically diverse (CALD) communities, older people, people of low socioeconomic status (SES), people in rural and remote communities, indigenous communities, and people with disabilities. It is often where these minority groups intersect that issues of access become more pronounced, such as for older NESB people. When considered together, accessibility cannot be deemed a "minority" issue (Leung 2014).

For most of the minority groups outlined in table 9.1, including settled refugees (who are grouped with new migrants), availability of technology is not an issue. There is ICT availability even in rural and remote areas, albeit limited, but uptake of computers and the Internet is low. Instead, telecommunications technologies have been privileged. The preference for mobiles to computers suggests that telecommunications offers a lower financial and technical threshold to participation.

However, Buckingham (2007) found that while some migrant groups were technologically well connected with friends and family, refugees were not. This suggests different accessibility issues at play between refugees and other migrants. Therefore, while there are commonalities in the ways minority groups experience the "digital divide," there are also differences in groups that are often deemed similar, such as migrants and refugees. These differences strengthen the case for accessibility and an intimate understanding of specific access needs to be able to design for all.

Table 9.1—Factors in minority group experiences with technology

	Refugees and new migrants	*NESB and CALD*	*Seniors*	*People with disability*	*Rural and regional*	*Remote indigenous*	*Low SES*
Availability is limited					✓	✓	
English literacy likely to be an issue	✓	✓		✓		✓	
Need for specialized equipment			✓	✓			
Prefer mobiles over computers	✓	✓	✓		✓	✓	✓
Affordability of telecommunication is an issue	✓	✓	✓	✓	✓	✓	✓
Lack technical literacies	✓	✓	✓	✓	✓	✓	✓
Experience isolation (social or geographic)	✓	✓	✓	✓	✓	✓	✓

Leung, 2014

HISTORIC LOCATION OF THE TERM "ACCESSIBILITY" WITHIN THE DISABILITY SECTOR

Although the notion of accessibility is informed by an ethic of inclusiveness, its focus has traditionally been in relation to disability, and designing technology products and services that consider the assistive devices that people with disabilities use to navigate the world.

In the past decade, accessibility has been embraced by web designers as a result of new policy and legislation, more accessible tools and general acceptance of this philosophical concept by web design professionals (Kennedy and Leung 2008: 70–71; Kennedy 2012). Firstly, there is the business argument: the number of people with disabilities (both physical and intellectual) is significant and will therefore always form a substantial proportion of end users. Secondly, regardless of the target market, there is the ethical principle of inclusiveness mentioned earlier, which stipulates that experiences, whether online or offline, should be accessible to everyone. Thirdly, there is the notion that "accessibility is a matter of usability" (Clarke 2006: 12). That is, accessible design is good design: "we should be designing our content so it is globally accessible and meets the needs of as many people as is possible and practical given our specific circumstances, regardless of their abilities or the type of device they choose to access the Web." (ibid.)

A social model of disability acknowledges that environments, whether physical, digital, or virtual, can enable or disable. A building that can only be entered via steps turns reliance on a wheelchair, zimmerframe, or pram into a disability, excluding those who use these technologies. A ramp, in lieu of stairs, exemplifies inclusive design that minimizes disability, as it does not adversely impact on wheelchair, zimmerframe, and pram users, nor those who do not rely on these technologies.

A social model asserts that it is society that is disabling, and by changing, adapting, and modifying society, and aspects of society, like buildings or the World Wide Web, to be inclusive, the world within which disabled communities live can become more enabling. Emerson et al. (2001: 19) sum up this position as follows:

> A social model of disability defines it as social restriction or oppression imposed by non-disabled others and advocates the removal of socially constructed barriers. By contrast, a medical model of disability defines disability as an individual problem, which needs to be "fixed."

The World Health Organization (2001) adopts the social model, recognizing that it is the environment, not the person, which is responsible for the difficulties experienced in having participation restricted or activities limited.

In other words, a social model places the onus on providers to make their products and services accessible, rather than compel users to have sufficient levels of linguistic, technical, or financial literacies to use them.

When applying social and medical models to technology use, the former contends that disability can be "designed out" by making technology experiences accessible to those who do not see, hear, move, communicate, or read well. A medical model would propose that increasing the digital literacy of people with disabilities would ensure they could access and participate in technology experiences on a par with the wider community.

As illustrated in table 9.3, people with disabilities are just one of many "minorities" to whom the social and medical models can be applied. Refugees are one of the other minority groups whose technology needs should be considered with accessibility in mind. Further, as will be explained in the remaining sections of this chapter, this may mean that the Internet is not necessarily the best vehicle to ensure refugee access to information and services.

APPLYING ACCESSIBILITY PRINCIPLES TO CONTENT

A user group such as refugees, while part of the broader umbrella of CALD communities, have specific needs that can be very different to other types of migrant and NESB community groups. Refugees typically choose and use technologies that are suited to their levels of language, technical, and financial literacy (Leung 2014).

Levels of literacy, both language and technical, are linked in that refugees who lack English language skills:

- often need assistance brokering access to technologies, thereby inhibiting their technical literacies;
- are unable to access or understand online content without help; and
- are also hindered in their ICT access because of confusing telecommunications contracts.

The predominance of textual content online makes the web inaccessible to many refugees with low levels of English literacy or those who are functionally illiterate, meaning they "cannot confidently read newspapers, follow a recipe, make sense of timetables, or understand the instructions on a medicine bottle" (ABS 2012). This inaccessibility is further exacerbated when users also have low levels of literacy in their native language. Not only is the content inaccessible for this user group, the technology itself—access to the Internet—is also largely inaccessible for those who have minimal exposure to or familiarity with computers. For respondent A37, having an email address is somewhat useless given her poor English language skills:

> Email address I have, but because I'm not good for reading and writing English, so I'm not using that. (A37)

The Internet is alienating for NESB consumers not only because content is overwhelmingly presented textually, but also due to the lack of available multilingual information. Therefore, access to services that are supposed to be universal—such as health and education—is impeded (NEDA 2010). This content impediment is manifested in many CALD/NESB consumers (such as refugees) simply choosing not to use the Internet. One study found that while they do not use the Internet to find health information, CALD communities are more open to using telecommunications for health care (AHWI 2012). This is exemplified in respondent A50 who does not use computers because she does not know how. However, she also says that phone cards are difficult to use and often do not work. Other respondents also said mobile phones were difficult to use because of everything being in English. As a result, respondent A53 did not know how to access her contacts or write a text message. She could only dial numbers and call people.

> You know when we are still new in the country we don't learn many things and we are not trained and we are not informed anywhere about how to use things. Yes lack of information. There are sometimes people from the telecommunication companies you can just make a contract over the phone. (A38)

It is unsurprising, then, that verbal communication was preferred over written forms. With voice calls being the main function used on the mobile phone, there was a tendency to trust, or possibly better understand, information that was disseminated by word-of-mouth over print or online content.

Nonetheless, the *Technology's Refuge* pilot study affirmed that with a combination of appropriate content provision and education in computer use, the Internet could become a more prominent information-seeking technology utilized by refugees (Leung, Finney Lamb, and Emrys 2009). A potential way of overcoming the generally low levels of English literacy among new refugee arrivals via Internet content provision is to use visually rich media as opposed to primarily text-based content. This concurs with accessibility design guidelines (Bywater 2005), which stipulate:

1. simplification and customization (allowing users to tailor and remove content unrelated to the task in hand);

2. multi-modality (allowing users to access content via the medium of their choice, be it textual, audio, or visual); and
3. interface choice (enabling content to work across platforms).

ACCESSIBILITY AS IT RELATES TO AFFORDABILITY

While much of the "digital divide" literature points to the marginalization of minority groups and communities, few studies specifically discuss affordability as a major factor, other than in studies of disability and ICT access. In the latter body of literature, the affordability of assistive technologies and their compatibility with computers and the Internet is discussed prominently (Dobransky and Hargittai 2006; Vicente and Lopez 2010; Macdonald and Clayton 2013).

According to Prasad (2013), the capacity to afford technology access is at the heart of issues of social exclusion as it is a requirement to function in society. For refugees, the challenge of affordability is directly connected to the lack of language, technical, and financial literacies. For example, there is a potential financial cost to not understanding contractual agreements with telecommunications and/or Internet service providers. There is also a cost to seeking help with technology access in terms of time, both for the person seeking assistance, as well as for the person assisting, otherwise called a "proxy user" (ACCAN 2016). There is strong evidence that refugees rely heavily on proxies to use technology or go online on their behalf. An Australian study illustrated the difficulties refugees face navigating the complex technical and contractual landscape of mobile phone deals, bundles and capped plans, especially with limited English language and native language literacies, demonstrating the unreasonable expectations of service providers that new migrants and refugees are vigilant and informed consumers in the local telecommunications market (Footscray Community Legal Centre 2011).

The difficulties surrounding low levels of financial literacy and affordability can begin to be addressed by aid agencies not only as part of the settlement process, but earlier still, during displacement. A refugee's ability to receive vital calls can be jeopardized by practical difficulties negotiating the lack of available telecommunications, or extant pay phone systems. Those without personal finances were especially disadvantaged in their ability to meet primary communication needs with the outside world. Respondents consistently referred to the prohibitive cost of a range of ICTs during their displacement.

> Many Arab people would have mobile phones, but Somalian people wouldn't have them, not for security reasons, but merely for the fact that they were

> too expensive. It was the same for landlines: they were safe to use (outside Somalia), but too expensive. (A53)

> Satellite telephone was also there but it was expensive and then unless you feel something burning, this is why you can go spend your money—it cost was expensive to use the satellite connection. (A7)

> At that time actually emails were available but I could not have access to the email because by that time it was not that easy to even own a computer. Although the Internet café they start coming out in say mid 90s where you can go and pay your money. Then I can remember sending an email to the UK, I wrote my email and then I take it to the post office and then they went and forwarded the email. So I remember like taking my message and then taking it to them so that they could type the email and then send it. (A7)

Respondent M8, who came through Turkey and Greece, claims to not have been able to make telephone calls because of the expense:

> No we were in shortage of money. I can't call and pay money for calling.

Although the Internet is often heralded as a cost-effective communication platform, in many countries of displacement, using Internet to communicate overseas was a more expensive option than the telephone for the recipients:

> I didn't contact my friend with the email in refugee camp [because it very expensive] so I just call with a phone. (I6 arr 2008)

Upon leaving for a settlement country, many refugees are saddled with financial obligation to family members still displaced or back in their country of origin (Lindley 2009). Many will promise to send whatever they can, whenever they can, to loved ones in refugee camps or communities they have left behind. This is necessary to help with survival in conflict areas, or to rebuild afterwards, or potentially to help others flee. Sperl (2001), for example, found that remittances from resettled refugees in Western countries to kin in Africa and the Middle East had become an important income source for the meeting of daily and critical needs, such as food, health care, housing, paying off debt, and sometimes education. Van Hear (2003) reported that global remittances by migrants (including refugees) were estimated to total US$100 billion annually, with 60 percent of this figure thought to be going to developing countries.

Refugees after settlement may also feel the need or pressure to sponsor their extended families to follow them to their settlement country. This will usually require the resettled refugee to bear the costs of mandatory medical examinations and travel expenses as part of their sponsored refugee's re-settlement application, as well as demonstrating sufficient financial resources to support their sponsored refugee(s) during their first weeks and/or months after arrival in the settlement country.

Riak (2005) has highlighted the importance of resettled refugees becoming self-sufficient as soon as possible by obtaining steady employment in their new country. However, when refugees do obtain paid employment, it tends to be within the lowest income bracket in their settlement country, meaning size-able portions of their income may be consumed by remittance obligations. Coupled with potentially high telecommunications costs of keeping in regular contact with loved ones left behind, this provides newly resettled refugees with additional sources of anxiety and stress in their new surroundings. While their priority is to maintain contact with family, the settlement process brings with it a whole set of new challenges, including how to navigate and under-stand how they are charged for their technology use.

Over a third of the thirty respondents in the *Mind the Gap* study had experienced unexpected costly bills. In some cases, they did not understand the nature of the expense. Others involved excess usage charges for the Internet. Settlement workers commented that clients often believed that what they had been charged was incorrect:

> Time to time actually I have people usually they're walk-in clients and they complain on the wrong bills, that they've been billed incorrectly for strange phone numbers or something like that or because again English issue . . . I think if just to generalise it probably the main issue is that they believe they were wrongly charged. (O2)

Another settlement worker described the case of a client whose home phone bill was disconnected after one month because of the amount owing on inter-national calls and calls to mobile phones.

Beyond the *Mind the Gap* study, further interviews with refugees showed similar experiences of bill shock. Respondent A48 waited one year after arriving in Australia in 1999 before purchasing a mobile phone. In the in-terim, she relied on her home phone. After receiving a very large home phone bill, she made the decision to have the home phone disconnected and use only her mobile phone instead. Even with help from her sponsor/proxy user to purchase the mobile phone and plan, she received a large bill, after which she stopped using the mobile.

Respondent A50 recounts almost exactly the same experience, except she was not able to pay her home phone bill and had the service disconnected by the provider. She blames her children for incurring the costs. After relying solely on the mobile phone, another large bill ensued and she stopped using it.

Respondents largely paid off the bills themselves without seeking further advice or assistance, and would devise their own strategies for minimizing spending on communication technologies. The challenges of maintaining telecommunications use meant that computers were not a priority. The cost of purchasing a computer and the ongoing costs of Internet access were prohibitive when compared with telecommunications technologies. Nor were computer and Internet access considered useful or meaningful in everyday life (Leung 2014): "Because if you are broke you can stay without Internet" (I10 arr 2009).

The *Mind the Gap* project called for policies regulating the conduct of the telecommunications industry to be strengthened in the areas of contracts and international phone cards. This is in order to protect consumers from refugee backgrounds with low levels of English language literacy from being locked into contracts which they do not fully understand, and which can subsequently lead to the development of unaffordable debts to telecommunications service providers (Leung 2011). Although affordability is a critical measure of accessibility, and can be regulated after the fact, the more preferable option is to have affordable access designed into the product or service in the first instance.

INSTRUMENTS OF ACCESSIBILITY

Technology's Refuge recommended that because technology is often financially inaccessible in refugee camps, cheap phone cards for calling overseas should be readily available. But is it possible for refugee service providers to guarantee access, not just availability? Given the diversity of the technological landscape, mechanisms are needed that are user-centered and inclusive of different types of access. In this sense, policy instruments may be useful to provide agencies with a clear mission and vision.

Given (2008: 92) regards the Universal Service Obligation (USO) as one of many instruments that can be used to ensure universal availability. Although originally intended for State telecommunications providers, a USO has primarily applied to making standard fixed line telephone services available and affordable. There is no equivalent USO for mobile telephone services in Australia, yet the inclusion of mobile and broadband services in the USO has already taken place in other countries (Prasad 2013: 227). Ironically, the mobile market has delivered near universal access in the absence of an USO, but does this mean it need not be governed by principles of inclusion? And could it work in situations of displacement, where network providers are compelled to ensure universal availability, while agencies that work directly with refugees are obliged to ensure accessibility?

A key principle underpinning the traditional USO may offer an initial departure point: to ensure "the ability of everyone to make and receive telephone calls at reasonable prices" (Blackman 1995: 171). A legal definition of USO implements this by requiring:

- Universal geographical availability
- Non-discriminatory access in terms of users and platforms
- Reasonable costs (ibid.: 172)

Kent (2007: 110) argues that a USO must go beyond geographical availability in order for access to be universal. Rather, all three aspects of a USO—the notions of "universality," "service," and "obligation"—each need to be interrogated separately in relation to access, affordability, and the current technological landscape. In order for a USO to have broad reach, Kent (ibid.: 117) recommends that it should be "ambiguous," high-level and principles-based, instead of tying it to particular services and funding models.

Other instruments, such as the Web Accessibility Initiative (WAI) and Web Content Accessibility Guidelines (WCAG), are also forms of USO policy in that they have been formulated to ensure that web content can be accessed through assistive devices by people with disabilities (W3C 2004). In relation to refugees' clear preference for phones over computers, such instruments can be used to ensure web content is mobile-friendly, if not, designed for mobile first.

This chapter has argued that technology accessibility should not be viewed only through the lens of the disability, as it traditionally has been. Simply making a technology available does not make it accessible, nor does it guarantee uptake. The experiences of refugees and technology that were outlined in the chapter demonstrated two critical factors informing accessibility: language and financial literacies.

If agencies are serious about providing technology access, and not just availability, then the attention must be paid to the ways in which content is presented to refugees, with respect to:

- designing information and services using language and content which are inclusive of the communities concerned and accessible to all;
- ensuring that access to information and services does not discriminate based on the technology used;
- accommodating an increasingly diverse technological landscape that encompasses old and new means of disseminating information.

Lack of affordability constitutes one of the greatest impediments to technology access and so needs to be framed as part of a wider discussion about inclusion. A rethinking of access and accessibility is required in which affordability is part and parcel of the design of inclusive technology products and services (Leung 2014).

The next chapter examines user-centered design and technology usability in relation to refugees. It will compare and contrast standard usability guidelines and analyze the challenges of testing and implementing these guidelines in situations of displacement. Assuming the hurdles to accessibility outlined in this chapter are overcome, technology products and services should always be designed to be as usable as possible.

Chapter Ten

User-Centered Design

User-centered design (UCD), also known as human-centered design or customer-centered design is premised upon understanding what users want from a system. What are the users' goals and what do they want to achieve? This necessitates gathering information from and about users including demographic data as well as insights into attitudes, behaviors and contexts of use (Cooper et al. 2007). UCD has been used across different disciplines such as in technology design, service design, product design, and experience design. Just as it can be applied in media environments to understand how users watch and interact with television, or business contexts to observe how customers move in and around a store, so too is it relevant to examining how refugees engage with the services of aid agencies in displacement contexts.

Lowgren and Stolterman (2004) refer to the design process as one that entails creating and shaping an artifact for use, and that this necessitates cooperating with users to ascertain their needs and expectations. Thus, the first stage of the UCD process is user research (AGIMO, no date). There are many methods for collecting data about users, but Cooper et al. (2007) contend that deep knowledge of users can only come from qualitative research techniques such as interviews, observation, and ethnography.

Furthermore, such methods are quicker and cheaper than using quantitative methods which involve large numbers of users.

Traditionally a method of anthropology, ethnography has been concerned with the detailed observation and description of a local system or community. It has been used to examine the social relations and rituals of language and its practice by a group of people, and how such communication processes construct identity. Soukup (2012: 74) explains ethnography as defining and/or discovering coherent boundaries of cultural identity and community.

Ethnography has increasingly found application in design practices as a method for gathering deep insights into customer or user needs and behaviors. Cooper and Dreher (2010: 41) argue that "ethnography provides perhaps the greatest insights into users' unmet and unarticulated needs, applications, and problems." Ethnography must acknowledge the fleeting and ephemeral nature of what it is studying, that Krizek (2007) likens to making meaning from pieces of a mosaic puzzle. The ethnographer's function is to weave together and interpret those "ethnographic fragments" (Gottschalk 1995: 195).

As a qualitative method, a level of interpretation is appropriate to penetrate beyond the superficial and get at something more "authentic" within the human experience (Collins 2010: 39). Therefore, it is important that ethnography, as a form of user research, is committed to "giving voice" as much as possible to users. This makes those who work directly with refugees on the ground well placed to act as ethnographers and user researchers, and as subject matter experts who can advocate for refugees as users of services.

From ethnographic research findings, it is then possible to develop a series of personas—representative clusters of actual users/consumers/clients who exhibit similar behavior patterns, feelings, and preferences (Cooper 2004). Good personas describe the key characteristics exhibited by a group of participants. Adlin and Pruitt (2010) outline that personas put "a face on the user—a memorable, engaging, and actionable image that serves as a design target." However, they are not every possible user nor do they replicate exactly specific clients (Brown 2006).

The findings from the refugee research in the *Mind the Gap* study informed the development of a series of fictitious but representative personas that would constitute the target audience for a consumer education program. The personas were used to ensure that the program would cater to the range of literacies across refugee communities.

USER PERSONAS

Ethra's Scenario of use

Ethra (figure 10.1) is at school in his IT class. It is in a computer lab. When the teacher isn't looking, his classmates are logging onto Facebook and messaging their friends. When he gets the chance, he does so too as it is free and doesn't cost any money, unlike doing it on his mobile. Anyway, he can't get Facebook on his phone right now as he has to get more credit for his phone. He knows his dad won't be happy with him as he asked for $30 for phone credit just a week ago. But he doesn't know how the credit ran out so quickly

Figure 10.1. Ethra.
Name: Ethra
Age: 19 years old
Ethnicity: Iraqi (Mandean)
Occupation: high school student
Location: Sydney, Australia
Language: Arabic (fluent), English (good)
Technology usage: internet (email, web especially Facebook) and mobile
iStockPhoto.com/Plougmann

and he can't explain to his dad either. He quickly gets Google up on the web as the teacher is nearby. He is not looking forward to doing his essay on IT networks, as he finds writing in English so hard after not going to school for three years while his family lived in Syria. He decides that once he gets more phone credit, he will text his friends to get together for a study group.

Fahima's Scenario of use

Fahima (figure 10.2) wishes she had more time to study and learn new technologies so she can keep up with her children, and they didn't have to translate so much for her. "I'm getting much better at English but it's still very embarrassing—they are so young, yet know so much of our business. I am quite ashamed." With five children to look after, she doesn't have a lot of opportunities to interact with others and practice her English. She looks forward to the times when she can call relatives in Egypt and Sudan, and

Figure 10.2. Fahima.
Name: Fahima
Age: 38 years old
Ethnicity: Sudanese
Occupation: mother of five
Location: Melbourne, Australia
Language: Arabic (fluent), English (intermediate)
iStockPhoto.com/oneclearvision

just speak freely in Arabic. It comforts her to hear their voices after losing touch with them for so long, not knowing their whereabouts after her family members had to scatter because of war in her home country. Even though it makes her feel at ease to be able to just pick up the phone at home and call, it costs a lot of money. She has heard that there are cheaper ways to call with special cards, but doesn't understand how these work and is not confident about listening to the instructions in English. What if she presses the wrong buttons? Anyway, she is ashamed to ask after receiving a very expensive phone bill. She had no idea that calling Sudan would cost so much money. She was even more surprised that calling numbers here in Australia can cost a lot of money too. Now she is worried about how to pay the bill and scared of making any more calls unless it is an emergency.

Aye's Scenario of use

"I miss most of school when I was in refugee camp so my English not good." Aye is glad that Australia accepted her and her family. They want to make a good life in Australia. She wants to get a good job but knows that for this she will need good English. Her family also want her to get a good job so she can help support them. She feels torn between doing more study and finding work so they can have more money. After all, they are waiting for her brother in Thailand to be accepted to come to Australia. And until he arrives, they have to send money to support him and other family members still in the refugee camp. They also have to call him from Australia because it costs too much for him to call from Thailand. She and her parents did their English classes when they got here, but her English is better than her parents, so they rely on her to help them when things need to be translated, to go to the shops to ask for things, to pay bills, and take them to the doctors and Centrelink. There is so much she has to do for them, she doesn't know how she will have time to study or work. Everything is new to her so it also takes time to learn. One of the new things she has had to learn is how to use the home phone. They have never had a phone in their home, but the settlement worker taught her how to put the numbers in. They had to find out from other Burmese people how to ring to Burma or Thailand. They told her to use a phone card, which you can buy at the shop. They showed her how to do it once, but it was hard to understand the English. Aye hopes that using the phone card won't make their home phone bill go up. A couple of times she tried the phone card and it didn't work, but when it does, it is good to be able to talk to her brother.

Personas are mainly utilized in conjunction with scenarios to narrate how users would interact with particular services or products. Scenario-based

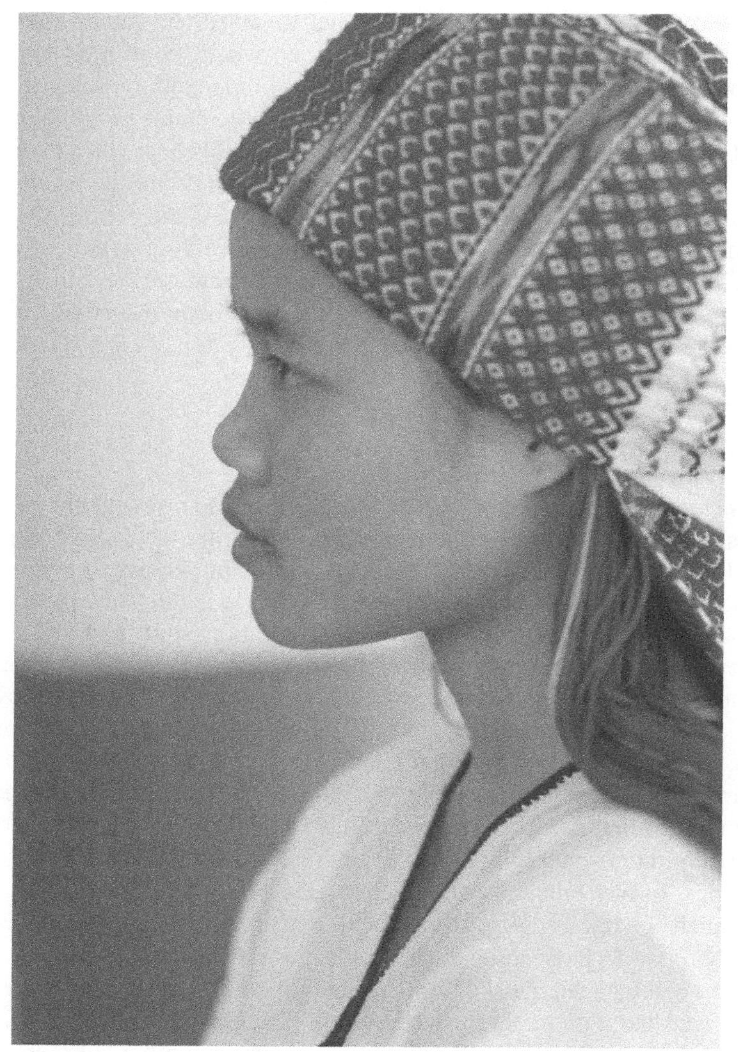

Figure 10.3. Aye.
Name: Aye
Age: 25 years old
Ethnicity: Burmese
Occupation: unemployed
Location: Brisbane, Australia
Language: Burmese (fluent), English (basic)
Technology usage: home phone and phone card
iStockPhoto.com/Juanmonino

design employs stories as ways of envisioning user goals. Visualization tools, such as storyboards allow the user story to be told with plot and brevity (Cooper et al. 2007).

In the *Mind the Gap* project, scenarios were used as a foundation for understanding the learning objectives of a training program that sought to assist newly arrived refugees navigate the technology landscape in Australia. An education consultant was engaged to design and develop a series of learning activities for the varying literacy levels of the user personas, as well as accommodate the different scenario contexts of each (ibid.).

Given the low levels of English language literacy among the respondents, the program prioritized oral rather than written methods of consumer education. That is, the program assumes face-to-face discussion among participants. Consumer education resource kits containing the learning activities were prototyped and tested by trainers, facilitators, volunteers, or caseworkers with refugee community groups. User testing was not only critical to improving the content of the kits, but also validating the authenticity of the personas and scenarios. Iterative testing is part of an ongoing cycle of UCD (Kuniavsky 2003).

Four units were prototyped for the consumer education program. These centered on the technologies for which refugees receive no formal training during settlement.

Unit 1: Landlines vs. mobile phones
Unit 2: Understanding mobile phones
Unit 3: Mobile phone plans
Unit 4: Phone cards

RESOURCE KIT UNITS

Unit 1: Landline vs. mobile phones

Lesson Objectives

Students will:

- Gain an understanding of the major differences between landline and mobile phones
- Become more familiar with terminology specifically related to phone technology
- Recognize the particular functions of landline and mobile phones
- Make more informed choices when purchasing a phone based on which type of phone best meets their needs

Resources:

- Phone type definition cards
- Definition word cards
- Label cards: mobile, versus, landline
- Landline and mobile phone photo cards
- Phone function cards
- Need cards
- Blu tac or tape

Teaching and Learning Activities:

Lesson Preparation: Divide a white/black board or wall space in half using chalk, marker or masking tape. Label one section "Landline," the other section "Mobile" and place the "versus" card on the dividing line.

1. Introduce this unit by reading the definitions of a landline and a mobile phone. Hand out the definition cards and have students place them under the correct heading: landline or mobile. As a group, they look at each definition card and discuss if it is placed in the right section.
2. Spread out phone photo cards on the floor or a desk so that all students have access. Ask students to sort cards into two piles: landlines and mobiles. Have students select photo cards and place them under the correct heading: landline or mobile. Again, ask the group to make sure that each photo is placed in the correct section.
3. Hand out phone function cards and ask students to read out their function. Discuss each function and explain further if needed. Ask students to place their function next to the matching phone type: landline or mobile. As a group, ensure that each function is in the right section.
4. Hand out need cards. One by one ask each student to display their card next to the phone type that best suits that need. If both phone types apply, then the card can be placed on the middle line.
5. Review the features of each phone type. Ask each student to state which phone is best suited to their needs.

Unit 2: Understanding mobile phones

Lesson Objectives

Students will:

- Gain an understanding of the major characteristics and functions of mobile phones

- Become more familiar with terminology specifically related to mobile phone technology
- Gain an understanding of the functions of a SIM card
- Become more familiar with different models of mobile phones

Resources:

- Examples of different models of mobile phones
- Examples of SIM cards
- Terminology cards and corresponding definition cards
- Phone and SIM card characteristic and function cards
- Butcher paper or white/black board
- Chalk or markers
- Blu tac

Teaching and Learning Activities:

1. Introduce this unit by handing out the mobile phones and allowing students to examine them and pass them around. Once the students have looked at several different examples, ask them to name the phone brands they have looked at or know about, write these on a board or butcher paper.
2. Hand out SIM card examples and allow students to pass around. Demonstrate how to insert a SIM card and then have students insert SIM cards into the phones.
3. Hand out characteristic and function cards. Ask each student to read out their card, then stick it on the board or butcher paper. Discuss and explain each card.
4. Give half the class a definition card and the other half a terminology card. Explain each one then ask the students to match each term to its corresponding definition. Display these on the board or butcher paper.
5. Review the phone models, characteristics, and functions of phones and SIM cards, terminology, and meanings. Allow for questions and discussion.

Unit 3: Mobile phone plans

Lesson Objectives

Students will:

- Become familiar with major mobile phone companies

- Gain an understanding of the major features of mobile phone plans and the fine print
- Become more familiar with terminology specifically related to mobile phone plans
- Be able to better understand a mobile phone bill
- Be able to make more informed choices regarding mobile phone plans

Resources:

- Mobile phone company logo cards
- Examples of different mobile phone plan advertisements
- Examples of different mobile phone bills with explanations of important sections
- Terminology cards and corresponding definition cards
- Phone plan comparison table
- Voting magnets or stickers
- Butcher paper or white/black board
- Chalk or markers and highlighters
- Blu tac

Teaching and Learning Activities:

1. Introduce this unit by asking students to identify the major mobile phone companies and display their logos on a board or wall.
2. Hand out examples of mobile phone plan advertisements while displaying the same examples on a board or wall. Have students highlight any terms in the ads that they do not understand or are unfamiliar. Ask students to read out their list of unknown words and highlight or circle on the display copies.
3. Using the list of unknown and unfamiliar terms given by students, match with definition cards and explain each one further if required.
4. Display enlarged copies of mobile phone bills. Deconstruct important sections of the bill by referring to the explanation cards. Allow for questions.
5. Create a table and compare phone plans by looking at main features. Have students vote for the best plan by using a magnet or sticker to indicate their choice.
6. Review the terminology and meanings, bill explanations, and different mobile phone plan advertisements. Allow for questions and discussion.

Unit 4: Phone cards

Lesson Objectives

Students will:

- Gain an understanding of phone card types, features, what they offer, and how to use them
- Become more familiar with terminology specifically related to phone cards
- Be able to make more informed choices regarding phone cards

Resources:

- Examples of different phone cards
- Terminology cards and corresponding definition cards
- Benefits and drawbacks cards
- Phone card comparison table
- Voting magnets or stickers
- Butcher paper or white/black board
- Chalk or markers
- Blu tac

Teaching and Learning Activities:

1. Introduce this unit by handing out examples of phone cards while displaying the same examples, enlarged, on a board or wall. Explain common features of cards using terminology and definition cards, for example, scratch pin numbers, access phone number, card amount, instructions.
2. Discuss benefits and drawbacks of phone cards, list on board or butcher paper.
3. Explain how to use cards, how to choose cards depending on customer requirements, and where to purchase.
4. Create a table and compare phone cards by looking at main features.
5. Review the card types, features, uses, and terminology. Allow for questions and discussion.

Testing for usability

The resource kits (figure 10.4) were tested with workers at a community legal center in Melbourne, clients of a refugee settlement support organization in Wollongong (south of Sydney), and young people from refugee backgrounds enrolled in an Intensive English Language Centre in school in southwest Sydney.

Key principles of usability (Nielsen 1995) formed the basis for testing. These included:

- *flexibility and efficiency of use*: could the learning activities be used in a variety of settings, and could they be undertaken quickly and easily?
- *user control and freedom*: did it allow the user a degree of autonomy over what and how they learned?
- *visibility of system status*: was it clear to the user what the activities were asking them to do?
- *consistency and standards*: were conventions adhered to across the four units?
- *error prevention and correction*: did it help users recognize, diagnose, and recover from errors?
- *match between system and real world*: were the learning activities authentic and useful in a meaningful way to the targeted user groups?

The feedback from these tests were as follows:

- Timing of this consumer education is crucial, as new refugee arrivals receive a mobile phone as part of their resettlement as soon as they get to

Figure 10.4. Prototype resource kits being tested (Leung, 2011)

Australia, but this is before they receive any of their compulsory English language tuition. Also, it is necessary to deliver this consumer education before critical decisions are made and contracts are signed on the basis of informal recommendations.

- Home phones would be irrelevant in some cases until consumers decide to install Internet at home. Even so, many are bypassing landlines and going straight to prepaid mobile Internet bundled with their mobile phone service provider.
- Given the critical role played by settlement organizations in orienting newly arrived refugees, the process of choosing telecommunications products and services perhaps should also be assisted by caseworkers and refugee support groups.
- Interpreters would also be needed at most workshops offered to recent arrivals who have not yet begun English classes.
- Experimentation with different formats for the resource kits is warranted so that reliance on written literacies is further minimized. Suggestions were made to utilize visual literacies more, that is, employ infographics, icons, and symbols where possible over text. Developing the resource kits around aural literacies in different languages was also worth considering given that the research showed respondents preferred this to reading and writing in English.
- It is redundant to attempt to simplify the key concepts of mobile phone plans, as they are often intentionally deceptive and misleading. The resource kit should also highlight the active misrepresentation by telecommunications providers and alert consumers to this, rather than adopt a neutral position on telecommunications products and services.

The applied practice of UCD to services designed for refugees does not necessarily have to incorporate technology. Whether or not technology is part of a solution depends entirely on the needs of user group(s) in a particular context. However, technology adds another layer of requirements that are not just technical. Irrespective of whether a service is no-tech, low-tech, or high-tech; simply making it available is not enough. It must also be accessible in ways that resonate with users' preferred platforms, modes of communication, and range of literacies.

Posters may be a low-tech method for communicating health messages, but would be challenging for refugee groups who cannot read or write in their native language.

Mobile SMS messaging to sustain connection with refugee communities also relies on content literacy (the ability to read and write), in addition to technical literacy (the ability to use the messaging function on mobile phones). More high-tech solutions such as video messaging or calling may

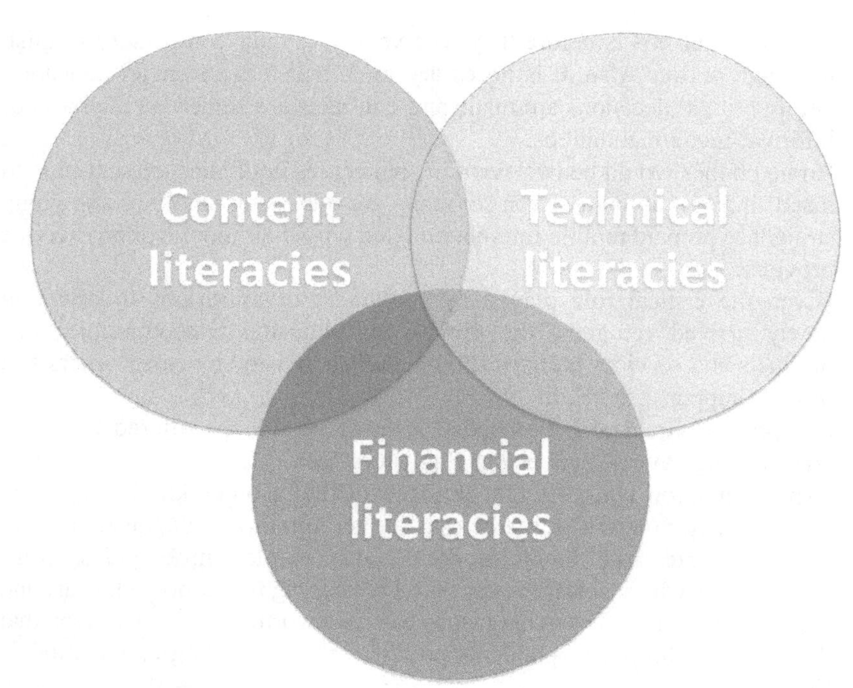

Figure 10.5. Digital literacies are constituted by a range of literacies.

not rely so much on content and technical literacies, but may be financially inaccessible to refugees if hardware or network access needs to be purchased. Contract terms for network access also require a level of financial literacy (understanding what one is paying for, and what is included in/excluded from the contract). UCD is committed to maximizing accessibility by tailoring experiences for the delicate balance between content, technical and financial literacies (figure 10.5) of its targeted users.

Chapter Eleven

UCD Principles in Practice

Many examples of user-centered design can be drawn from refugees themselves, especially those who have set up businesses in refugee camps such as Za'atari in Jordan. There are numerous cases of successful refugee enterprises:

> Mohammed Jendi owns the biggest store in Zaatari, a large clothing emporium. His main tip is to work hard to know your customers. The way he went about this would impress any management consultant: he explains how he surveyed his friends and neighbours to find out what they needed and what they wanted. At the start, when conditions were really desperate, they just needed clothing to make it through the harsh winter. But as things improved, the refugees wanted a more individual look. So Jendi now offers men an array of colourful tracksuits, sports jackets and jeans in various cuts. Women can choose from a huge selection of shawls, handbags and high heels. (Davies 2016)

In another case, a woman named Rose who runs a pharmacy and health clinic in the Kyangwali refugee camp in Uganda, found her business was being affected by the inconsistent electricity supply to the settlement. She decided to purchase her own generator to power her own business, but also to be able to sell electricity to other households and businesses in the camp affected by poor power supply. In solving her own problem, Rose is able to turn the solution into an enterprise that addresses the same issue for others (Betts et al. 2015: 12).

Refugee entrepreneurship articulates the first rule of usability: know thy user. Their businesses succeed because they understand the needs of refugees from personal experience. When family and friends in Za'atari found that they were located on opposite sides of the camp and wanted to move to be closer to one another, refugee entrepreneurs devised their own way of relocating their allocated temporary housing. Using the fence posts demarcating the camp's boundaries, they designed and built trailers that enabled the housing units to be easily moved around the camp, and developed a removalist business that allowed refugees to shape the environment and architecture of Za'atari according to their needs (Bloom 2016). These are not only examples of user-centered design, but innovation as "dynamic problem-solving among friends" (Betts et al. 2015: 4).

Beyond knowing the needs and wants of refugees in a particular context and designing for those requirements, others argue for a much more participatory approach, that is, involving refugees in the design and development process:

> When designing technologies for underserved populations, it's simply not enough to design for the user. You need to design *with* the user . . . This approach, often called human-centred design, design thinking, or participatory design, starts from the assumption that solutions need to be created *with* the people they are designed to serve, not *for* them . . . Embedding yourself with the communities and people you intend to help is not simply good practice, it's essential. (Levinson 2016)

This kind of "bottom-up," grassroots or participatory innovation has been missing from technology solutions to refugee problems (Bloom 2016). Thus, it is important to note that in addition to being "bottom-up," user-centered design can also be "top-down," imposed from above or by others onto refugees, although this carries the risk of assumptions being made about the affected community and proposed solutions subsequently not being adopted. Somewhere in between are projects where agencies consult with the local population and then prototype and test a possible solution. The application of Boda Boda Talk Talk model is an example of this, whereby new arrivals to the Bidibidi refugee settlement in Uganda wanted information about the services available to them. Discussions with refugees on the ground by the UNHCR Emergency Lab found that there was a desire for information to be disseminated orally or in audio form, due to low literacy levels. The agency then used an existing model, pioneered by Internews and borrowed from the Juba refugee settlement in South Sudan, in which motorbikes kitted out with audio equipment ride around the settlement playing short radio shows at different "listening points." The sites also become places where refugees gather to talk and share information (Drew 2016).

Such "in-between" forms of user-centered design that mix together "top-down" and "bottom-up" methods can also be seen in open innovation platforms. Primarily, open innovation is a model whereby organizations use external and internal knowledge to innovate (Chesbrough 2004). It is innovation that is enabled through distributed sources of knowledge (Bessant et al. 2014), is decentralized in nature (Murray and O'Mahony 2007), and focused on receiving external feedback and proposals (Benkler 2006; von Hippel 2005). With the advent of the Internet, there has been significant growth in the use of open innovation in the private sector (Lafley and Charan 2008). The rapid increase in efficient and low-cost computerized communication has enabled open innovation to develop naturally for many firms (Chesbrough 2006; Benkler 2006). In relation to the refugee/humanitarian/development sectors, open innovation is characterized by:

1. Crowdsourcing: Afuah and Tucci (2012: 355) define crowdsourcing as "the act of outsourcing a task to a crowd, rather than to a designated "agent" . . . such as a contractor, in the form of an open call" as a way of improving the efficiency and effectiveness of problem solving. Crowdsourcing is often used to solve dilemmas faced by an organization, by drawing on the collective expertise of those who may be currently working in or have previously worked in the field; refugees themselves; or anyone who believes that they can constructively contribute ideas, knowledge or possible solutions to an identified problem. In this sense, it combines input from the top, bottom and everything in between to develop a user-centered solution.
2. Competition: by setting a time-limited challenge with incentives for participation, creating a sense of competition in the urgent pursuit of solutions to a problem also serves to quickly mobilize people into a temporary self-governing community that ensures an end result is produced in a timely and meaningful manner, but with mechanisms that encourage experimentation and play (O'Mahony and Ferraro 2007).
3. Open source: open source is particularly focused on providing the external user with access to the core knowledge that forms the collective asset that is being improved.

An example of an open innovation platform using crowdsourced competition is OpenIDEO, a community-driven website aimed at bringing individuals together to solve global issues. On the platform are challenges that are active for three to five months.

Challenges can be sponsored by an organization, and the platform allows participants to propose ideas as well as to provide feedback on and show support for ideas. Winning ideas may receive funding to be prototyped and tested. A refugee education challenge, sponsored by the UNHCR and UNICEF, posed the question "How might we improve education and expand learning opportunities for refugees around the world?" It generated 399 research contributions, 376 ideas of which 26 were refined, 7 were voted the top ideas, and 5 went on to be funded (OpenIdeo 2016). As Boudreau and Lakhani (2009: 70) write:

> These disparate examples illustrate how participants can learn from and build upon the discoveries of others by "standing on the shoulders of giants"—in which the "giant" is collective knowledge. In such innovation initiatives, the community participants work with technologies or components that are closely related, thereby creating a foundation for subsequent efforts.

Therefore, it allows a range of solutions to be designed and developed simultaneously, which can be customized for specific contexts. One example of this is the public availability of all the interview data collected in my research on refugees and technology through the online database at http://trr.digimatter.com.

Such data may be used by agencies as a form of user research on refugees' technology literacies so that service providers can avoid having to collect similar data themselves. The sharing of this data means organizations can potentially bypass an entire stage of the user-centered design process and can focus on design, development and testing of prototypes.

Conclusion

Back in 1992, the United Nations High Commissioner for Refugees, Sadako Ogata, lamented that the relationship between refugees and the environment has long been overlooked (Harper 2016). Indeed, the same can be said about the relationship between refugees and technology. Despite the massive changes in global environmental health and the technology landscape, refugee projects addressing either of these issues are still deemed "luxuries to be implemented only when more urgent matters are attended to." The longitudinal data in the book drawn from hundreds of surveys and interviews with refugees demonstrates that while technologies change, the ideas surrounding technology use and users persist over time.

In particular, notions of digital divides which are understood only in binary and mutually exclusive terms pervade technology debate, discussion and policy. Such binaries include the concept of a singular "digital divide" with a Netizen majority on the "right" side, and minority groups like refugees on the "wrong" side. However, now just as at other critical times in history, the danger of enduring binary concepts like Us and Them is apparent. Closer examination reveals that there are a series of dichotomies that are geographic, socioeconomic, and far from clearcut. There is no majority versus minority. Instead, there are theoretical dualities, binary models deployed against humanity. Techno-dystopian and utopian fantasies are deeply embedded in Australia's system of mandatory detention of asylum seekers and other numerous technology projects aimed at refugees but which do not profoundly engage with their needs.

Nonbinary models that analyze the relationship between technology and its users disrupt the narrative of a singular "digital divide," and allow for greater complexity. Theories of the strength of weak ties, Actor Networks, and hierarchies of technology literacies can be applied and adapted for diverse environments without marginalizing refugees as nonusers or "victims" of digital exclusion. Such models lay the foundations for practical human-centered approaches to designing technology products, platforms, and services for refugee contexts.

According to Greifinger (in Cheney 2016), the practice of user-centered design is often at odds with how things are done in traditional development settings. Ultimately, there is no "one way" of doing UCD. Rather, it should be considered a principles-based practice that is underpinned by an ethic of inclusion and access for all. The notion of accessibility as distinct from availability is key, as is the acknowledgment that access is multifaceted: linguistic, technical, and financial. The principle of being user-centered and usable is also critical, with the needs of the user(s) being at the center of all design decisions. At its most extreme, participatory design has the user(s) making those decisions and leading the design process. Finally, participation is central to the practice of open innovation, which seeks to crowdsource ideas, designs and solutions to problems from users, experts, and anyone connected to the issue at hand.

The examples provided in this book not only illustrate the aforementioned theories and principles in practice. They also point to the vast potential for new ways of thinking and doing in displacement contexts when refugees are conceptualized differently: as technology users, as active agents, and networkers, and as resourceful innovators.

Bibliography

Adlin, T. and Pruitt, J. (2010) *The persona lifecycle: Your guide to building and using personas*. Burlington: Elsevier.

Afuah, A. and Tucci, C. (2012) "Crowdsourcing as a solution to distant search," *Academy of Management Review*, 37(3): 355–375.

Ajana, B. (2013) "Asylum, identity management and biometric control," *Journal of Refugee Studies*, 26(4): 576–595.

Ajana, B. (2013a) *Governing through Biometrics: The Biopolitics of Identity*. Basingstoke: Palgrave Macmillan.

Alexander, D. (2014) "Social media in disaster risk reduction and crisis management," *Science and Engineering Ethics*, 20(3): 717–733.

Alizadeh, A. (2012) "The Ogre" in Rundle, S. and Bharat, M. (eds) *Alien Shores: Tales of Refugees and Asylum Seekers from Australian and the Indian Subcontinent*. Melbourne: Brass Monkey Books.

Amnesty International (2016) "Island of Despair: Australia's 'Processing' of Refugees on Nauru." Online 17 October 2016, available at: https://www.amnesty.org.au/wp-content/uploads/2016/10/ISLAND-OF-DESPAIR-FINAL.pdf

Anderson, B. (2006) *Imagined Communities: Reflections on the origin and spread of nationalism*. New York: Verso.

Anderson, J. (2013) "Policy report on UNHCR's Community Technology Access Program: Best practices and lessons learned," *Refuge*, 29(1): 21–29.

Ang, I. (1996) *Living room wars: Rethinking media audiences for a postmodern world*. London: Routledge.

Appadurai, A. (1996) *Modernity at large: Cultural dimensions of globalization*. Minneapolis: University of Minnesota Press.

The Asylum Seeker and Refugee Law Project (2013) "What is mandatory detention," *The Asylum Seeker and Refugee Law Project*. Online 29 December 2017, available at: https://uqrefugeeresearch.wordpress.com/2013/07/08/what-is-mandatory-detention/

Australian Bureau of Statistics (ABS) (2012) "National year of reading: Libraries helping make Australia a nation of readers," *1301.0—Yearbook Australia, 2012*. Online 21 January 2016, available at: http://abs.gov.au

Australian Communications and Media Authority (ACMA) (2015) *Australians' Digital Lives*. Online 1 January 2016, available at: http://acma.gov.au/theACMA/Library/Corporate-library/Corporate- publications/communications-report

Australian Communications Consumer Action Network (ACCAN) (2016) "Going online on behalf of others," *ACCAN magazine*, 8(Summer): 17.

Australian Council of Heads of Schools of Social Work (ACHSSW) (2006) *"We've boundless plains to share": The first report of the People's Inquiry into detention*. Melbourne: Australian Council of Heads of Schools of Social Work.

Australian Government (2015) *Communications report 2013–2014 series: Report 1—Australian digital lives*. Canberra: Australian Communication and Media Authority.

Australian Government Information Management Office (AGIMO) (no date) User Profiling and Testing Toolkit. Online 5 October 2016, available at: https://www.finance.gov.au/sites/default/files/User-Profiling-and-Testing-Toolkit.pdf

Australian Health Workforce Institute (AHWI) (2012) *Telecommunications and health information for multicultural Australia*. Sydney: Australian Communications Consumer Action Network.

Ayanso, A., Cho, D., and Lertwachara, K. (2014) "Information and communications technology development and the digital divide: a global and regional assessment," *Information Technology For Development*, 20(1): 60–77.

Baker Jr, H. (1991) "Hybridity, the rap race and pedagogy for the 1990s," in Penley, C. and Ross, A. (eds) *Technoculture*. Minneapolis: University of Minnesota Press.

Baldassar, L., Baldock, C., and Wilding, R. (2007) *Families caring across borders, migration, ageing and transnational caregiving*. Hampshire: Palgrave Macmillan.

Barnard, Y., Bradley, M., Hodgson, F., and Lloyd, A. (2013) "Learning to use new technologies by older adults: Perceived difficulties, experimentation behaviour and usability," *Computers in Human Behaviour*, 29(4): 1715–1724.

Barzilai-Nahon, K. (2006) "Gaps and bits: conceptualizing measurements for digital divide/s," *The Information Society*, 22(5): 269–278.

Benkler, Y. (2006) *The wealth of networks: How social production transforms markets and freedom*. New Haven, CT: Yale University Press.

Bernal, V. (2006) "Diaspora, cyberspace and political imagination: The Eritrean diaspora online," *Global Networks*, 6(2): 161–179.

Bessant, J., Ramalingam, B., Rush, H., Marshall, N., Hoffman, K., and Gray, B. (2014) "Innovation management, innovation ecosystems and humanitarian innovation," UK Department for International Development. Online June 2014, available at: http://r4d.dfid.gov.uk/pdf/outputs/Hum_Response/Humanitarian-Innovation-Ecosystem-research-litrev.pdf

Betts, A., Bloom, L., and Weaver, N. (2015) *Refugee Innovation: Humanitarian innovation that starts with communities*. Oxford: Humanitarian Innovation Project, University of Oxford.

Blackman, C. (1995) "Universal service: Obligation or opportunity," *Telecommunications Policy*, 19(3): 171–176.

Bloom, L. (2016) "5 ways to better engage with bottom-up innovations by refugees." Online 12 January 2016, available at: http://innovation.unhcr.org/5-ways-to-better-engage-with-bottom-up-innovations-by- refugees/

Bolter, J. and Grusin, R. (1999) *Remediation: Understanding new media*. Cambridge, MA: MIT Press.

Boudreau, K. and Lakhani, K. (2009) "How to manage outside innovation," *MIT Sloan Management Review*, 50(4): 69–76.

Brady, F. and Dyson, L. (2014) "Enrolling mobiles at Kowanyama: Upping the ANT in a remote Aboriginal community," *International Conference on Culture, Technology, Communication*, Oslo: University of Oslo, 179–194.

Briskman, L. (2013) "Technology, control and surveillance in Australia's immigration detention centres," *Refuge*, 29(1): 9–19.

Briskman, L., Latham, S. and Goddard, C. (2008) *Human rights overboard: Seeking asylum in Australia*. Melbourne: Scribe.

Broadbent, R. and Papadopoulos, T. (2013) "Bridging the digital divide—an Australian story," *Behaviour and Information Technology*, 32(1): 4–13.

Brown, D. (2006) *Communicating Design*. Berkeley: New Riders.

Bruno, G.; Esposito, E.; Genovese, A. and Gwebu, K. (2011) "A critical analysis of current indexes for digital divide measurement," *Information Society*, 27(1): 16–28.

Buckingham, D. (2007) "Digital media literacies: Rethinking media education in the age of the Internet," *Research in Comparative and International Education*, 2(1): 43–55.

Bywater, B. (2005) "Accessibility guidelines can be simple," UsabilityNews.com. Online 21 January 2016, available at: http://usabilitynews.bcs.org/content/conWebDoc/47463

Castan Centre for Human Rights Law (2003) "Detention, children and asylum seekers: A comparative study," *Submission to the National Inquiry into Children in Immigration Detention*. Online 31 January 2017, available at: https://www.monash.edu/law/research/centres/castancentre/our-areas-of-work/refugees-and-asylum-seekers/policy/child

Charmarkeh, H. (2013) "Social media usage, tahriib (migration) and settlement among Somali refugees in France," *Refuge*, 29(1): 43–52.

Cheney, C. (2016) "How to develop a human-centered design mindset," *Devex*. Online 30 October 2016, available at https://www.devex.com/news/how-to-develop-a-human-centered-design-mindset-89006

Chesbrough, H. (2004) "Managing open innovation," *Research Technology Management*, 47(1): 23–26.

Chesbrough, H. (2006) *Open innovation: The new imperative for creating and profiting from technology*. Boston: Harvard Business School.

Children Out of Immigration Detention (2017) *ChilOut Newsletter: December 2017 edition*. Online 23 December 2017, available at: http://www.chilout.org/news_chilout_newsletter_december_2017

Clarke, A. (2006) *Transcending CSS: The fine art of web design*. Berkeley, CA: New Riders.

Collins, H. (2010) *Creative research: The theory and practice of research for the creative industries*. London: Thames and Hudson.

Coombes, B. (2009) "Generation Y: Are they really digital natives or more like digital refugees?," *Synergy*, 7(1): 31–40.

Cooper, A. (2004) *The Inmates are Running the Asylum: Why high-tech products drive us crazy and how to restore the sanity*. Indianapolis: Sams Publishing.

Cooper, A.; Reimann, R. and Cronin, D. (2007) *About Face: The essentials of interaction design*. Indianapolis: Wiley Publishing.

Cooper, R. and Dreher, A. (2010) "Voice-of-customer methods: What is the best source of new product ideas?" *Stage-Gate International*, Winter: 38–48.

Cranny-Francis, A. (2005) *Multimedia*. London: Sage Publications.

Cunningham, S. (2001) "Popular media as public 'sphericules' for diasporic communities," *International Journal of Cultural Studies*, 4(2): 131–147.

Currion, P. (2011) "Technology: Bringing solutions or disruptions?" *Forced Migration Review*, 38(October): 41–42.

Curtin, J. (2001) "A digital divide in rural and regional Australia?" *Current Issues Brief 1 2001–02*. Online 2 September 2016, available at: http://www.aph.gov.au/About_Parliament/Parliamentary_Departments/Parliamentary_Library/Publications_Archive/CIB/cib0102/02CIB01

Danielson, N. (2013) "Channels of protection communication, technology and asylum in Cairo, Egypt," *Refuge*, 29(1): 31–42.

Dankova, P. and Giner, C. (2011) "Technology in aid of learning for isolated refugees," *Forced Migration Review*, 38(October): 11–12.

Davies, R. (2016) "Business tips from a refugee camp." Online 8 December 2016, available at: https://www.1843magazine.com/dispatches/the-daily/business-tips-from-a-refugee- camp

De Leeuw, S. and Rydin, I. (2007) "Migrant children's digital stories: Identity formation and self-representation through media production," *European Journal of Cultural Studies*, 10(4): 447–468.

Del Prete, A.; Calleja, C. and Cervera, M. (2011) "Overcoming generational segregation in ICTs: Reflections on digital literacy workshop as a method," *Gender Technology and Development*, 15(1): 159–174.

Delanty, G. (2002) "Two conceptions of cultural citizenship: A review of recent literature on culture and citizenship," *The Global Review of Ethnopolitics*, 1(3) March.

Delgado-Moreira, J. (1997) "Cultural citizenship and the creation of European identity," *Electronic Journal of Sociology*, 2(3).

Dewan, S. and Riggins, F. (2005) 'The digital divide: Current and future research directions,' *Journal of the Association for Information Systems*, 6(12): 298–337.

Diawara, M. (1993) *Black American Cinema*. New York: Routledge.

Diken, B. (2004) "From refugee camps to gated communities: Biopolitics and the end of the city," *Citizenship Studies*, 8(1): 83–106.

Dobransky, K. and Hargittai, E. (2006) "The disability divide in internet access and use," *Information, Communication and Society*, 9(3): 313–334.

Drew, K. (2016) "Innovation or imitation in Uganda." Online 15 November 2016, available at: http://innovation.unhcr.org/innovation-imitation-in-uganda/

Duale, A. (2011) "How displaced communities use technology to access financial services," *Forced Migration Review*, 38(October): 28–29.

Eastin, M., Cicchirillo, V. and Mabry, A (2015) "Extending the digital divide conversation: Examining the knowledge gap through media expectancies," *Journal of Broadcasting and Electronic Media*, 59(3): 416–437.

Emerson, E., Hatton, C., Flece, D. and Murphy, G. (2001) *Learning disabilities: The fundamental facts*. London: The Foundation for People with Learning Disabilities.

Ennis, L., Rose, D., Denis, M., Pandit, N. and Wykes, T. (2012) "Can't surf, won't surf: The digital divide in mental health," *Journal of Mental Health*, 21(4): 395–403.

Epstein, D., Nisbet, E., and Gillespie, T. (2011) "Who's responsible for the digital divide? Public perceptions and policy implications," *Information Society*, 27(2): 92–104.

Fiske, L. (2016) "Human rights and refugee protest against immigration detention: Refugees' struggles for recognition as human," *Refuge*, 32(1): 18–27.

Fiske, L. (2016a) *Human Rights, Refugee Protest and Immigration Detention*. London: Palgrave Macmillan.

Footscray Community Legal Centre. (2011). *Taking Advantage of Disadvantage: case studies of refugee and new migrant experiences in the communications market*. Sydney: Australian Communications Consumer Action Network.

Gajjala, R. (6 August 1999) "Cyborg Diaspora and Virtual Imagined Community: Studying SAWNET," *Cybersociology Issue 6: Research Methodology Online*.

Gifford, S. and Wilding, R. (2013) "Digital escapes? ICTs, settlement and belonging among Karen youth in Melbourne, Australia," *Journal of Refugee Studies*, 26(4): 558–575.

Gillespie, M. (1995) *Television, ethnicity and cultural change*. London: Routledge.

Gilroy, P. (1993) *The Black Atlantic: Modernity and double consciousness*. Verso: London.

Given, J. (2008). "The eclipse of the universal service obligation: Taking broadband to Australians." *Journal of Policy, Regulation and Strategy for Telecommunications, Information and Media*, 10(5/6): 92–106.

Glazebrook, D. (2004) "Becoming mobile after detention," *Social Analysis: International Journal of Cultural and Social Practice*, 48(3): 40–58.

Global Detention Project (2008) "Australia Immigration Detention." Online 29 December 2017, available at: https://www.globaldetentionproject.org/countries/asia-pacific/australia

Gottschalk, S. (1995) "Ethnographic fragments in postmodern spaces," *Journal of Contemporary Ethnography*, 24(2): 195–228.

Graham, J. (2014) *Video games: Parents' perceptions, role of social media and effects on behaviour*. Hauppage: Nova Science Publishers.

Graham,M. and Khosravi, S. (2002) "Reordering public and private in Iranian cyberspace: Identity, politics and mobilization," *Identities: Global Studies in Culture and Power*, 9(2): 219–246.

Granovetter, M. (1983) "The strength of weak ties," *Sociological Theory*, 1: 201–233.

Gurak, L. (2001) *Cyberliteracy: Navigating the Internet with awareness*. New Haven: Yale University Press.

Hailovich, H. (2013) "Bosnian Austrians: Accidental migrants in trans-local and cyberspaces" *Journal of Refugee Studies*, 26(4): 524–540.

Hale, T.; Cotten,S.; Drentea,P. and Goldner,M. (2010) "Rural-urban differences in general and health-related Internet use," *American Behavioral Scientist*, 53(9): 1304–1325.

Hall, M. (2011) "The only constant is change," *Forced Migration Review*, 38(October): 9–11.

Hall, S. (1998) "Aspirations and attitude . . . Reflections on black Britain in the 90's" *New Formations: Frontlines, Backyards*, 33: 38–46.

Halleck, D. (1991) "Watch out Dick Tracy! Popular video in the wake of Exxon Valdez," in Penley, C. and Ross, A. (eds) *Technoculture*. Minneapolis: University of Minnesota Press.

Harney, N. (2013) "Precarity, affect and problem solving with mobile phones by asylum seekers, refugees and migrants in Naples, Italy," *Journal of Refugee Studies*, 26(4): 541–557.

Harper, A. (2016) "A critical time for refugees and their environment (again)." Online 10 December 2015, available at http://innovation.unhcr.org/critical-time-refugees-environment/

Henwood, F., Wyatt, S., Miller, N., and Senker, P. (2000) "Critical perspectives of technologies, in/equalities and the information society," in Wyatt, S., Henwood, F., Miller, N., and Senker, P. (eds) *Technology and In/equality*. London: Routledge.

Horst, H. (2006) "The blessings and burdens of communication: Cell phones in Jamaican transnational social fields," *Global Networks*, 6(2): 143–159.

Howard, E. and Owens, C. (2002) "Using the Internet to communicate with immigrant/refugee communities about health." Poster presentation at ACM/IEEE Joint Conference on Digital Libraries, Portland, Oregon, 14–18 July.

Huggins, R. and Izushi, H. (2002) "The digital divide and ICT learning in rural communities: examples of good practice service delivery," *Local Economy*, 17(2): 111–122.

Human Rights and Equal Opportunity Commission (HREOC) (2004) "A last resort? National inquiry into children in immigration detention." Online 31 January 2017, available at: http://www.humanrights.gov.au/publications/last-resort-national-inquiry-children-immigration-detention

Human Rights Law Centre (2015) "International community condemns Australia's treatment of asylum seekers during major human rights review at UN." Online 10 November 2015, available at: https://www.hrlc.org.au/news/international-community-condemns-australias-treatment-of-asylum-seekers-during-major-human-rights-review-at-un

Human Rights Watch (2016) *World Report 2016: Events of 2015*. Online 29 December 2017, available at: https://www.hrw.org/sites/default/files/world_report_download/wr2016_web.pdf

International Telecommunications Union (2013) "Global Internet Usage." Online 31 January 2017, available at: http://en.wikipedia.org/wiki/Global_Internet_usage

International Telecommunication Union (ITU) (2015) *Measuring the Information Society Report 2015*. Geneva: ITU.

Internet Live Stats (2016) "Internet users." Online 31 August 2016, available at: http://www.internetlivestats.com/internet-users/

Isaacs, M. (2014) *The Undesirables: Inside Nauru*. Melbourne: Hardie Grant.

Jaivin, L. (2012) "Karim," in Rundle,S. and Bharat,M. (eds) *Alien Shores: Tales of Refugees and Asylum Seekers from Australian and the Indian Subcontinent*. Melbourne: Brass Monkey Books.

Jakubowicz, A. (2016) "European leaders taking cues from Australia on asylum seeker policies." Online 7 November 2016, available at: https://theconversation.com/european-leaders-taking-cues-from-australia-on-asylum-seeker-policies-66336

Jung, Y.; Peng, W.; Moran, M; Jin, S.; McLaughlin, M.; Cody, M.; Jordan-Marsh, M.; Albright, J. and Silverstein, M. (2010) "Low-income minority seniors enrollment in a cybercafe: Psychological barriers to crossing the digital divide," *Educational Gerontology*, 36(3): 193–212.

Kabbar, E. and Crump, B. (2006) "The factors that influence adoption of ICTs by recent refugee immigrants to New Zealand," *Informing Science Journal*, 9(9): 111– 121.

Kadende-Kaiser, R. (2000) "Interpreting language and cultural discourse: Internet communication amongst Burundians in the diaspora," *Africa Today*, 47(2): 120– 148.

Karim, K. (2003) *The media of diaspora*. London: Routledge.

Kaspersen, A. and Lindsey, C. (2016) "The digital transformation of the humanitarian sector." Online 5 December 2016, available at: http://blogs.icrc.org/law-and-policy/2016/12/05/digital-transformation-humanitarian-sector/

Kennedy, H. (2012) *Net Work: Ethics and Values in Web Design*. Basingstoke: Palgrave Macmillan.

Kennedy, H. and Leung, L. (2008) "Lessons from web accessibility and intellectual disability," in Leung, L. (ed), *Digital Experience Design: Ideas, Industries, Interaction*. Bristol: Intellect.

Kent, M. (2007) "New technology and the universal service obligation in Australia: drifting towards exclusion?," *Nebula*, 4(3): 111–123.

Kluzer, S. and Rissola, G. (2009) "E-Inclusion policies and initiatives in support of employability of migrants and ethnic minorities in Europe," *Information Technologies and International Development*, 5(2) Summer: 67–76.

Krizek, R. (2007) "Ethnography as postmodern practice: Accessing meaning in the ephemeral moments of events," *National Communication Association Annual Convention—Communicating Worldviews: Faith-Intellect-Ethics*. Chicago, November.

Kuniavsky, M. (2003) *Observing the user experience: A practitioner's guide to user research*. Waltham: Morgan Kaufmann Publishers.

Kushner, T. and Knox, K. (1999) *Refugees in an age of genocide: Global, national and local perspectives during the twentieth century*. London: Taylor and Francis.

Lafley, A. and Charan, R. (2008) *The game changer*. New York: Profile.

Latour, B. (1992) "Where are the missing masses? The sociology of a few mundane artifacts" in Bijker, W. and Law, J. (eds) *Shaping Technology/Building Society: Studies in Sociotechnical Change*. Cambridge, MA: MIT Press.

Latour, B. (1999) "On recalling ANT" in Law, J. and Hansard, J. (eds) *Actor network theory and after*. Oxford: Blackwell.

Law, J. (1999) "After ANT: complexity, naming and topology" in Law, J. and Hansard, J. (eds) *Actor network theory and after*. Oxford: Blackwell.

Lems, A., Gifford, S., and Wilding, R. (2016) "New myths of OZ: The Australian beach and the negotiation of national belonging by refugee background youth," *Continuum: Journal of Media and Cultural Studies*, 30(1): 32–44.

Lenhart, A. (2010) *Teens and mobile phones*. Washington, DC: Pew Research Center.

Lenhart, A. and Horrigan, J. (2003) "Re-visualizing the digital divide as a digital spectrum," *IT and Society*, 1(5): 23–39.

Leung, L. (2011) *Mind the Gap: Refugees and communications technology literacy*. Sydney: Australian Communications Consumer Action Network. Online 20 November 2011, available at: http://accan.org.au/files/Reports/Mind the Gap ACCAN-UTS.pdf

Leung, L. (2014) "Availability, access and affordability across 'digital divides:' Common experiences across minority groups," *Australian Journal of Telecommunications and the Digital Economy*, 2(2). Online 7 February 2017, available at: http://doi.org/10.7790/ajtde.v2n2.38

Leung, L. and Finney Lamb, C. (2010) *Refugees and Communication Technology*. Sydney: UTS Shopfront. Online 4 December 2017, available at: https://www.uts.edu.au/partners-and-community/initiatives/uts-shopfront-community-program/news-and-events/news-0/refugees

Leung ,L., Finney Lamb, C. and Emrys, L. (2009) *Technology's refuge: The use of technology by asylum seekers and refugees*. Sydney: UTS Shopfront.

Levinson, A. (2016) "Making an app to help refugees? Read this first," *Responsible Business*. Online 5 May 2016, available at: http://prosper.community/making-an-app-to-help-refugees-read-this-first/

Lindley, A. (2009) "The early morning phonecall: Remittances from a refugee diaspora perspective," *Journal of Ethnic and Migration Studies*, 35(8): 1315–1334.

Livingstone, S. and Helsper, E. (2007) "Gradations in digital inclusion: Children, young people and the digital divide," *New Media and Society*, 9(4): 671–696.

Livingstone, S., Van Couvering, E., and Thumim, N. (2005) *Adult media literacy: A review of the research literature*. London: Ofcom.

Loader, B., Vromen, A., and Xenos, M. (2014) *The networked young citizen: Social media, political participation and civic engagement*. London: Routledge.

Lowgren ,J. and Stolterman, E. (2004) *Thoughtful interaction design*. Cambridge, MA: The MIT Press.

Luster, T., Qin, D., Bates, L., Johnson, D., and Rana, M. (2009) "The lost boys of Sudan: Ambiguous loss, search for family, and reestablishing relationships with family members," *Family Relations*, 57(4): 444–456.

MacCallum, M. (2002) "Girt by sea: Australia, the refugees, and the politics of fear," *Quarterly Essay*, 5: 1–73.

Macdonald, S. and Clayton, J. (2013) "Back to the future, disability and the digital divide," *Disability and Society*, 28(5): 702–718.

Madianou, M. and Miller, D. (2012) "Polymedia: towards a new theory of digital media in interpersonal communication," *International Journal of Cultural Studies*, 16(2): 169–187.

Mallapragada, M. (2000) "The Indian diaspora in the USA and around the web," in Gauntlett, D. (ed.) *Web studies: Rewiring media studies for the digital age*. London: Arnold.

Mares, P. (2002) *Borderline: Australia's response to refugees and asylum seekers in the wake of the Tampa*. Sydney: UNSW Press.

Martinez-Pecino, R. and Lera, M. (2012) "Active seniors and mobile phone interaction," *Social behaviour and Personality*, 40(5): 875–880.

McIver, W. and Prokosch, A. (2002) "Towards a critical approach to examining the digital divide," *IEEE monograph*. Piscataway, New Jersey.

McLeod, S. (2012) *What school leaders need to know about digital technologies and social media*. San Francisco: Jolley-Bass.

McMaster, D. (2002) *Asylum seekers: Australia's response to refugees*. Melbourne: Melbourne University Press.

Melkote, S. and Liu, D. (2000) "The role of internet in forging a pluralistic integration," *Gazette*, 62(6): 495–504.

Migliorino, P. (2011) "Digital technologies can unite but can also divide: CALD communities in the digital age," *Australian Public Libraries Information Service*, 24(3): 107–110.

Miniwatts Marketing Group (2014) "Internet world stats: Usage and population statistics." Online 31 January 2017, available at: http://www.internetworldstats.com/stats.htm

Mitra, A. (1997) "Virtual commonality: Looking for India on the Internet," in Jones,S. (ed) *Virtual culture: Identity and communication in cybersociety*. London: Sage Publications.

Mitzner, T., Boron, J., Fausset, C.; Adams, A., Charness, N., Czaia, S., Dijkstra, K., Fisk, A., Rogers, W., and Sharit, J. (2010) "Older adults talk technology: Technology usage and attitudes," *Computers in Human Behaviour*, 26(6): 1710–1721.

Morris, J., Mueller, J., and Jones, M. (2014) "Use of social media during public emergencies by people with disabilities," *Western Journal of Emergency Medicine*, 15(5): 567–574.

Mulgan, G. (1996) "High tech and high angst" in Dunant, S. and Porter, R. (eds) *The age of anxiety*. London: Virago.

Murray, F. and O'Mahony, S. (2007) "Exploring the foundations of cumulative innovation: Implications for organization science," *Organization Science*, 18: 1006–1021.

Naidoo, S. and Raju, J. (2012) "Impact of the digital divide on information literacy training in a higher education context," *South African Journal Of Libraries and Information Science*, 78(1): 34–44.

Nathan, N. and Zeitzer, J. (2013) "A survey study of the association between mobile phone use and daytime sleepiness in California high school students," *BMC*

Public Health, (13): 840. Online 29 September 2014, available at: http://www.biomedcentral.com/1471K2458/13/840

National Ethnic Disability Alliance (NEDA) (2010) *Communicating difference: Understanding communications consumers from non-English speaking backgrounds*. Sydney: Australian Communications Consumer Action Network.

Nielsen, J. (1995) "10 usability heuristics for user interface design." Online 5 October 2016, available at: https://www.nngroup.com/articles/ten-usability-heuristics/

Norman, D. (2004) *Emotional design: Why we love (or hate) everyday things*. New York: Basic Books.

O'Mahony, S. and Ferraro, F. (2007) "The emergence of governance in an open source community," *Academy of Management Journal*, 50: 1079–1106.

O'Mara, B. and Harris, A. (2016) "Intercultural crossings in a digital age: ICT pathways with migrant and refugee-background youth," *Race, Ethnicity and Education*, 19(3): 639–658.

Omata, N. (2011) "Online connection for remittances," *Forced Migration Review*, 38(October): 27–28.

OpenIdeo (2016) "Challenge: How might we improve education and expand learning opportunities for refugees around the world?." Online 5 February 2016, available at: https://challenges.openideo.com/challenge/refugee-education/funded

Parham, A. (2004) "Diaspora, community and communication: Internet use in transnational Haiti," *Global Networks*, 4(2): 199–217.

Phillips, C. (2013) "Remote telephone interpretation in medical consultations with refugees: Meta-communications about care, survival and selfhood," *Journal of Refugee Studies*, 26(4): 505–523.

Prasad, R. (2013) "Universal service obligation in the age of broadband," *The Information Society: An International Journal*, 29(4): 227–233.

Price, G. and Richardson, L. (2011) "Access to information—inclusive or exclusive?," *Forced Migration Review*, 38(October): 14–15.

Reid, M. (1993) *Redefining Black film*. University of California Press, Oxford.

Riak, S. (2005) "Remittances as unforeseen burdens: the livelihoods and social obligations of Sudanese refugees," *Global Migration Perspectives*, 18. Geneva: Global Commission on International Migration.

Robertson, Z., Wilding, R., and Gifford, S. (2016) "Mediating the family imaginary: Young people negotiating absence in transnational refugee families," *Global Networks*, 16(2): 219–236.

Rosaldo, R. (1994) Cultural citizenship in San Jose, California, *PoLAR*, 17(2): 57–63.

Ruffer, G. (2011) "What Ushahidi can do to track displacement," *Forced Migration Review*, 38(October): 25–26.

Sein, M. and Furuholt, B. (2012) "Intermediaries: Bridges across the digital divide," *Information Technology for Development*, 8(4): 332–344.

Selwyn, N. (2004) "The information aged: A qualitative study of older adults' use of information and communications technology," *Journal of Aging Studies*, 18: 369–384.

Soukup, C. (2012) "The postmodern ethnographic flaneur and the study of hypermediated everyday life," *Journal of Contemporary Ethnography*, 42(2): 226–254.

Sparks, C. (2013) "What is the 'Digital Divide' and why is it important?" *Javnost—The Public*, 20(2): 27–46.

Sperl, S. (2001) *Evaluation of UNHCR's policy on refugees in urban areas: a case study review of Cairo*. Geneva: UNHCR Evaluation and Policy Analysis Unit.

Suwamaru, J. and Anderson, P. (2012) "Closing the digital divide in Papua New Guinea: A proposal for a national telecommunications model," *Contemporary PNG Studies*, 17(Nov): 1–15.

Townsend, L., Sathiaseelan, A., Fairhurst, G., and Wallace, C. (2013) "Enhanced broadband access as a solution to the social and economic problems of the rural digital divide," *Local Economy*, 28(6): 580–595.

Underwood, J. (2008) "Varieties of Actor-Network Theory in information systems research," *European Conference on Research Methods*, London, 1–7.

UNHCR (2010) "Convention and protocol relating to the status of refugees." Online 22 December 2017, available at: http://www.unhcr.org/en-au/3b66c2aa10

UNHCR (2013) "UNHCR statistical online population database," Online 29 May 2014, available at: http://www.unhcr.org/statistics/populationdatabase

UNHCR (2016) "UNHCR mid-year trends 2015." Online 26 December 2017, available at: http://www.unhcr.org/56701b969.html

United Nations (1948) "Universal declaration of human rights." Online 1 February 2017, available at: http://www.un.org/en/universal-declaration-human-rights/

Van Deursen, A. and van Dijk, J. (2014) "The digital divide shifts to differences in usage," *New Media and Society*, 16(3): 507–526.

Van Hear, N. (2003) "Refugee diasporas, remittances, development and conflict," *Migration Information Source*. Washington DC: Migration Policy Institute. Online 21 January 2016, available at: http://www.migrationinformation.org/feature/display.cfm?ID=125

Vertovec, S. (2004) "Cheap calls: The social glue of migrant transnationalism," *Global Networks*, 4(2): 219–224.

Vicente, M. and Lopez, A. (2010) "A multidimensional analysis of the disability digital divide: Some evidence for Internet use," *The Information Society*, 26: 48–64.

Vigdor, J., Ladd, H., and Martinez, E. (2014) "Scaling the digital divide: home computer technology and student achievement," *Economic Inquiry*, 52(3): 1103–1119.

Villaveces, J. (2011) "Disaster Response 2.0," *Forced Migration Review*, 38(October): 7–9.

Vinson, T. and Rawsthorne, M. (2015) *Dropping off the Edge 2015: persistent communal disadvantage in Australia*. Richmond and Curtin: Jesuit Social Services and Catholic Social Services Australia.

von Hippel, E. (2005) *Democratizing Innovation*. Cambridge, MA: MIT Press.

Wall, I. (2011) "Citizen initiatives in Haiti," *Forced Migration Review*, 38(October): 4–6.

Walsh, S.; White, K. and Young, R. (2008) "Over-connected? A qualitiative exploration of the relationship between Australian youth and their mobile phones," *Journal of Adolescence*, 31: 77–92.

Warschauer, M. (2002) "Reconceptualizing the digital divide," *First Monday*, 7(7), Online 22 September 2016, available at: http://firstmonday.org/ojs/index.php/fm/article/view/967/888

Warschauer, M. (2003) *Technology and Social Inclusion: Rethinking the Digital Divide*. Cambridge, MA: MIT Press.

Wilding, R and Gifford, S. (2013) "Introduction," *Journal of Refugee Studies*, 26(4): 495–504.

Wilding, R. (2006) "'Virtual' intimacies? Families communicating across transnational contexts," *Global Networks*, 6(2): 125–142.

Williams, B. (2001) "Black secret technology: Detroit techno in the information age," in Nelson, A. and Tu, T. (eds) *Technicolor: Race, technology and everyday life*. New York: New York University Press.

Wolfinbarger, S. and Wyndham, J. (2011) "Remote visual evidence of displacement," *Forced Migration Review*, 38(October): 20–21.

World Café Community Foundation (2015) "A Quick Reference Guide for Hosting World Café." Online 31st January 2017, available at: http://www.theworldcafe.com/wpGcontent/uploads/2015/07/Café-To-Go-Revised.pdf

World Health Organization (2001) *International classification of functioning, disability and health*. Online 21 January 2016, available at: http://www.who.int/classifications/icf/en/

World Wide Web Consortium (W3C) (2004) *Web accessibility initiative*. Online 21 January 2016, available at: https://www.w3.org/WAI/

World Wide Web Consortium (W3C) (2004) *Web content accessibilty guidelines 2.0*. Online 21 January 2016, available at: https://www.w3.org/TR/WCAG20/

Index

access, 8, 19, 45, 91
accessibility, 5, 93, 94, 97, 103
Actor Network Theory, 4, 47, 59, 60
affordability, 8, 33, 100, 103, 104
asylum seeker(s), 5, 25, 26, 28
aural literacies, 83, 90
Australia
  Migration Act, 28, 32
  refugee program, 6
Australian Communications Consumer Action Network
  funded *Mind The Gap* study, 14
availability, 8, 11, 21, 45, 91, 94, 95, 103

Christmas Island, 53
closed camps, 29, 30
consumer education resource kits, 113
content literacy, 119
cultural citizenship, 26

database of refugees and technology, 124
  trr.digimatter.com, 11
digital divide(s), 19, 23, 27
  "haves", "have nots", 20
  geographic, 20
  statistics, 18
  studies, 17
digital exclusion, 18
digital literacy, 18, 20, 89, 94
  training, 21
disability, 97
  medical model, 97
  social model, 97
dystopianism, 4, 37, 43, 62

emotional design, 37
ethnography, 107, 108

financial literacy, 94, 100
financial obligation, 101

Granovetter
  strength of weak ties, 47, 49, 50, 53, 54, 57
  theory of weak ties, 4

immigration detention, *See also* mandatory detention, 25, 53
  technologies, 25
immigration detention centers, 36
intermediate countries, 77
internally displaced persons (IDPs), 5

laggards, 19
literacy
  financial, 98

language, 98
technical, 98

mandatory detention, *See* also immigration detention, 25
Manus Island, 53
Migration Act, 54
minority groups, 95
mobile citizenships *See also* cultural citizenship, 26
mobile phone, 9

Nauru, 25, 53
Netizen, 23, 25, 26
networks, 34
Noula, 44

offshore processing, 53
open detention, 29, 33
open innovation, 122

perimeter technologies, 40
personas *See also* user personas, 40
persons of concern, 5, 6
pilot study, 12
polymedia, 8

refugee camps, 50, 53, 55, 60, 74, 76, 121
refugee convention, 6
refugee entrepreneurship, 121
Refugee Studies, 11
refugee(s), 5, 6
definition, xi
technology users, 7
remittances, 101

social determinism, 4, 23, 35
social model of disability, 97
socio-technical networks, 49, 52
Socio-Technical Studies (STS), 4
stateless persons, 5

technical literacy, 119
technologically advantaged, 19
technological determinism, 4, 5, 23, 35, 43
technologies of representation, 38
technologies of surveillance, 37
technology access, 94
technology literacy, *See also* digital literacy, 65
Technology Studies, 4 *See also* Socio-Technical Studies (STS)
technology users, 8
telecommunications literacy, 65, 73, 78
theory of weak ties, 4

unauthorized arrivals, 28
United Nations (UN) Refugee Agency, xi
United Nations High Commissioner for Refugees (UNHCR), xi, 5, 6, 8, 45, 125
Universal Declaration of Human Rights, 44
Universal Service Obligation, 103
unlawful noncitizen, 28
usability, 118
user research, 108
user scenarios, 108
user-centered design, 5, 107, 121, 126
Ushahidi, 44
utopianism, 4, 37, 42, 62

verification technologies, 40

Web Accessibility Initiative, 104
Web Content Accessibility Guidelines, 104

About the Author

Linda Leung is an associate professor and honorary associate at the University of Technology Sydney, where she has directed postgraduate studies in digital innovation and transformation, creative and cultural industries, and interactive media. She has previously taught and/or conducted research at the universities of London, East London, North London, Miami, and Western Sydney.

She has also authored *Digital Experience Design: Ideas, Industries, Interaction,* which chronicles the diverse backgrounds of practitioners in the dot.com world, and subsequently, the theories, ideas, models, and frameworks they bring and apply to the design of technologically mediated experiences. Her first book, *Virtual Ethnicity: Race, Resistance and the World Wide Web,* is concerned with how technology is appropriated by those with limited access to it, as well as the problems and possibilities which arise when technology is made available to minority groups. It draws from the disciplines of technology studies, media/communication studies, and anthropology/cultural studies. This cross-disciplinary approach also informs her teaching and research on digital creative industries, project management processes and practices, and user experience design.

CPSIA information can be obtained
at www.ICGtesting.com
Printed in the USA
BVHW04*1317140718
521314BV00009B/11/P